T3-BSK-161

The New Optoelectronics
Ball Game

IEEE PRESS
445 Hoes Lane, PO Box 1331
Piscataway, NJ 08855-1331

1992 Editorial Board
William Perkins, *Editor in Chief*

K. K. Agarwal	K. Hess	A. C. Schell
R. S. Blicq	J. D. Irwin	L. Shaw
R. C. Dorf	A. Michel	M. Simaan
D. M. Etter	E. K. Miller	Y. Sunahara
J. J. Farrell III	J. M. F. Moura	J. W. Woods
	J. G. Nagle	

Dudley R. Kay, *Executive Editor*
Carrie Briggs, *Administrative Assistant*

Karen G. Miller, *Assistant Editor*

Technical Reviewers

Arpad A. Bergh
Bellcore

James E. Brittain
Georgia Institute of Technology

Steward Flaschen
Oxbridge Partners

Alan McAdams
Cornell University

Serafin Menocal
Bellcore

The New Optoelectronics Ball Game

The Policy Struggle
Between the U.S. and Japan
for the Competitive Edge

Philip Seidenberg

IEEE
PRESS

The Institute of Electrical and Electronics Engineers, Inc., New York

This book may be purchased at a discount from the publisher
when ordered in bulk quantities. For more information contact:

IEEE PRESS Marketing
Attn: Special Sales
PO Box 1331
445 Hoes Lane
Piscataway, NJ 08855-1331
Fax: (908) 981-8062

The views expressed in this book are solely those of the author and
not those of the IEEE, Inc. The IEEE PRESS serves as an
independent book publishing division of the IEEE and is in no way
connected to the United States Activities Board, Standards, or other
policy-making divisions. This book is published as a service to the
technological community at large as a reference point for discussion
and exchange. Responsibility for its contents rests upon the author
and not upon the IEEE, its Societies, nor its members.

©1992 by the Institute of Electrical and Electronics Engineers, Inc.
345 East 47th Street, New York, NY 10017-2394

*All rights reserved. No part of this book may be reproduced in any form, nor
may it be stored in a retrieval system or transmitted in any form, without
written permission from the publisher.*

Printed in the United States of America

10 9 8 7 6 5 4 3 2 1

ISBN 0-7803-0406-3
IEEE Order Number: PP0301-2

Library of Congress Cataloging-in-Publication Data

Seidenberg, Philip N.
    The new optoelectronics ball game : the policy
struggle between the U.S. and Japan for the
competitive edge / by Philip N. Seidenberg
        p.   cm.
    "IEEE order number: PP0301-2"—Verso t.p.
    Includes bibliographical references (p.   ) and
index.
    ISBN 0-7803-0406-3
    1. Optoelectronics industry—Government policy—United States.
2. Optoelectronics industry—Government policy—Japan. I. Title.
HD9696.0673U67   1992
338.4'762138152'0973—dc20   91-44004
                                                              CIP

Contents

Preface

Most policy books on semiconductors to date have been written by people not directly connected with the industry. Economists, social scientists, and political scientists are the first to note the societal changes wrought by technological innovations. They analyze the public and private sector policies that have affected these changes, identify problem areas, and recommend courses of action. Their greatest contribution is in the role of an early warning system. However, when the observers and analysts from these "outsider" disciplines switch from policy analysis to policy design, the ensuing recommendations are too broad-based and often off the mark. They tend to mix problems with answers, and are unable to precisely craft a doable policy. Even within the electronics industry, many fail to understand why the U.S. has lost its semiconductor leadership. They are trapped within their particular technological paradigm, or blinded by self-interest.

Most of the recommendations on how to get back on the semiconductor track are as generic as the Ten Commandments.

Many recommendations are not doable because of political or economic interest groups whose views on national industrial policies conflict, resulting in a policy design stalemate.

In this book, I shall discuss the semiconductor, optoelectronics, and the optoelectronic integrated circuit (OEIC), how they are related, why they are critical to U.S. industrial competitiveness, and how the U.S. can improve its competitive position through the manufacture of the OEIC. I shall describe the status of optoelectronic device technology in Japan and the U.S., and demonstrate that the U.S. lags behind Japan in the development of this technology. The optoelectronic integrated circuit, conceived and developed by the U.S., is a semiconductor device that can give the U.S. an opportunity to restore its competitiveness.

I shall make specific policy recommendations that can be implemented speedily and economically. I do not favor policies that require massive changes in the U.S. social, political, or cultural structures. It is better to proceed in small steps when dealing with bureaucracies. In a direct confrontation, innovation rarely defeats bureaucracy.

In Chapter 1, I cite the concern for the U.S. loss of semiconductor competitiveness. A brief definition of the semiconductor and optoelectronics is given. Major technological innovations are listed. I show that U.S. semiconductor policy analysts have not understood the importance of optoelectronic device technology. This has intensified the loss of U.S. competitiveness in electronics. I identify the OEIC as a key device in the battle for leadership in computers and communications.

Chapter 2 explains the technology of optoelectronic semiconductors. The scope and size of the market are defined. The importance of the OEIC is described.

Chapter 3 lists the semiconductor firms in the U.S. and Japan that are developing OEICs. It shows that the top Japanese firms are committed to optoelectronics, whereas the leading merchant semiconductor firms in the U.S. remain uncommitted.

Chapter 4 estimates the amount of money spent by the governments and semiconductor firms of the U.S. and Japan for optoelectronic research and development (R&D). Japan started committing resources to optoelectronic R&D earlier than the U.S.

Chapter 5 analyzes the number and content of technical papers on OEICs in five leading journals. The analysis from these indicators reveals that Japan is leading in the commercial development of OEIC technology.

Chapter 6 describes the optoelectronic policies of Japan. It demonstrates how well organized the Japanese were in identifying and targeting optoelectronic technology and the OEIC. Japan's assiduousness in acquiring this technology and its strategy of dominating support technologies have been major factors in its competitive success in optoelectronics.

Chapter 7 reveals that the U.S. has no coherent national policy on optoelectronics. Government reports are cited to show the conflicting policies of the United States. Consortia in the U.S. are a reaction to contemporary market conditions rather than long-range technological planning.

In Chapter 8, I recommend specific policies that the U.S. should adopt in order to be competitive in OEICs. They include a national OEIC foundry, government procurement of OEICs from domestic sources only, export decontrol, and a technology database. These are minimum steps designed not to unduly upset the political power structure.

This book includes an Appendix and a References section. The Appendix contains statistical information that is useful, but not integral to the main text. A source is acknowledged in the text by a parenthetical citation with that information necessary to identify it in the References section, usually the author's name and page numbers, and sometimes the date when a last name is referenced more than once.

Acknowledgments

I am indebted to John Havick for helping me make the transition from the somewhat intuitive and loosely structured logic of the entrepreneur to the tightly ordered logic of the scholar. His guidance was invaluable in the orderly presentation of my research. I thank John Endicott whose expertise on Japan has reinforced my foundation of knowledge. To John McIntyre, I owe thanks for providing me with background material on Japanese management philosophies. I am grateful to James Brittain and Steward Flaschen whose encouragement and advice contributed to the publication of this book.

I give thanks with love to my wife, Katherine, whose wholehearted support made the research and writing of the book more agreeable, and to Cassandra, my daughter, who had the unenviable task of typing the manuscript.

1

The Problem

The U.S. is losing its international competitiveness. International competitiveness is defined as:

> the ability and willingness of firms to identify, adopt and pursue strategies leading to their long-term preeminence over rivals in an industry. It involves appropriate strategies for decision-making, marketing, design, technical effort, investment and manufacturing, and firm structure and manpower [Sciberras and Payne, p. viii].

A 1989 study by the Massachusetts Institute of Technology (M.I.T.) Commission on Industrial Productivity is scathing in its indictment of U.S. industry for the deterioration of its manufacturing capability and the decline of its electronics sector [Weber, pp. 55–58]. Some of the members of the commission viewed industrial America in macroeconomic terms to determine if there are any patterns to its industrial decline [Berger *et al.*, pp. 39–47]. Six weaknesses were observed:

- outdated strategies;
- neglect of human resources;

- failures of cooperation;
- technological weaknesses in development and production;
- government and industry working at cross-purposes; and
- short time horizons.

The Chairman and President of the Council on Competitiveness and Chief Executive Officer (CEO) of the Hewlett Packard Company considers U.S. government policy outdated because it does not recognize the importance of promoting commercial applications of technology ["Council: More money to U.S. industry," p. 26]. The National Research Council believes that the Federal Government plays an insufficient role in "assessing technological opportunities and catalyzing development of technology in industry" [National Research Council, p. 68, 1988]. The U.S. Activities Board of the Institute of Electrical and Electronics Engineers (IEEE) is concerned with the inadequate support "for the continuing education of engineers" [IEEE–USA, p. 15]. The CEO of Harman International Industries and a former Undersecretary of the Department of Congress notes that labor and management do not cooperate in U.S. factories, and that workers are not encouraged to be creative [Rosenthal, p. B-1]. The President of Bell Communications Research (Bellcore) bemoans the U.S. weakness in telecommunications, attributing it to the drive for short-term profit and the lack of a "coordinated national strategy for technology" ["Bellcore chief hits drive for fast profits, policy lack," p. 6]. Two Professors at Stanford University believe that the U.S. semiconductor industry "needs to improve its process skills" in high-volume production [Okimoto and Rowen, p. 14].

The U.S. loss of competitiveness is most serious in the semiconductor industry. The semiconductor is the technology driver for electronics. It is not possible for the U.S. to maintain its competitive edge in communications and computers if it irreversibly loses its historical leadership in such a fundamental technology. Semiconductor technology is constantly evolving as device innovations are introduced periodically.

The goal of this book is to demonstrate how U.S. industrial competitiveness can be enhanced by reversing the decline of the U.S. semiconductor industry. This decline will not be arrested by chasing after device markets and technologies in which leadership already has been lost. The U.S. must identify and exploit emerging semiconductor technologies as a way to restore its market dominance in electronics and other high-tech industries.

The focus of this work is on a major device innovation, the optoelectronic integrated circuit (OEIC). The OEIC is a semiconductor device which will greatly contribute to improved system performance, cost, and reliability as computers and communications merge their functions. Fundamentally, the OEIC is an integrated circuit (IC) which processes signals electrically, and transmits these signals optically. This is in contrast to the typical IC, which both processes and transmits signals electrically. Although the OEIC is in laboratory development, its introduction into the marketplace is not far off.

The Japanese understood much earlier than did the Americans the technological importance of merging light and electronic functions. Although a U.S. researcher first conceived this idea [Miller, pp. 2059–2069, 1969], and another U.S. researcher developed the first OEIC in 1978 [Lee *et al.*, p. 806], the U.S. semiconductor industry did not recognize the market importance of the OEIC until the late 1980's. In 1979, within one year of its development, the Japanese had identified the OEIC as a driver technology in communications, and had initiated a national program to develop the device [Hayashi *et al.*, p. 1431].

The U.S. now is waking up to the importance of the optical functions in electronic systems. According to a recent Gallup poll of U.S. engineers, optical integrated circuits rank second in a priority list of 18 technologies [Rosenblatt, pp. 22–27]. This relatively narrow technology ranked ahead of superconductivity, nuclear fusion, intelligent robots, artificial organs, and a host of other well-publicized technologies. The engineers, evenly divided among academia, government, and industry, considered only the full use of natural energy resources as a more important national priority than the optical integrated circuit.

The survey, sponsored by the *Nihon Keizai Shimbun*, a leading Japanese business daily, akin to the *Wall Street Journal*, asked these members of the Institute of Electrical and Elec-

tronics Engineers (IEEE), the world's largest professional engineering organization, to rank the most important factors in
promoting U.S. dominance in science and technology in the
21st century. The responses ranked industrial research and development (R&D) first, followed by stronger government pro-
motion of advanced technology using Japan as a model, and
more federal funding for the development of science and technology.

In 1991, a former Deputy Undersecretary of Defense, now
the Vice President of Technology Assessment for Boeing Aero-
space Company, wrote: "the path to photonic progress . . . the
need for low-cost high functionality optoelectronic integrated
circuits (OEICs)" [Martin, p. 85].

A leading U.S. technology research and consulting firm
forecasts that in five–eight years, OEICs will leverage a computer market of $20 billion, in three–five years a cable television
market of $5 billion, and in five–ten years a switching system
market of $10 billion, all total, $35 billion [Fulenwider, p. 40].

This book will show why the OEIC is an important semiconductor device, and that semiconductor technology is critical
for U.S. industrial competitiveness. It will explain how the merging of the integrated circuit and the optoelectronic semiconductor has resulted in the OEIC. The importance of the OEIC
as an enabling technology for communications and electronics
will be demonstrated. U.S. and Japanese policies in the identification, targeting, funding, and developing of the OEIC will
be reviewed. Finally, some recommendations will be offered
which the U.S. can initiate to be competitive in the manufacture
and marketing of the OEIC.

The Semiconductor

The role of the semiconductor in transforming the social and
economic fabric of the world during the last half of the 20th
century is well established. The semiconductor, created out of
the needs of the communications industry, has become the most
pervasive technical innovation since World War II. It is the core
of all high-speed electronic systems designed to sense, process,

store, transmit, and display information. The semiconductor is the key enabling technology in the control and measurement of industries ranging the entire industrial gamut from high technology to smokestack. Even the highest of the high-technology sectors, as defined by the U.S. Chamber of Commerce ["Technology's top ten," p. 8], including such nonelectronic areas as inorganic chemicals, rubber and synthetic fibers, drug and medicines, engines, and plastics, all depend upon semiconductor-based electronic sensing, instrumentation, and data processing equipment for the design and manufacture of their products.

A semiconductor is a material whose electrical conductivity at room temperatures is superior to an insulator (ceramic) and inferior to a conductor (copper). It tends to behave as an insulator at temperatures approaching absolute zero, and as a conductor as temperatures rise above room temperature.

The eight major semiconductor device innovations to date, save one, are of U.S. origin. They are listed below with date of disclosure and reference.

1947	Transistor	[Bardeen, p. 69]
1957	Tunnel Diode	[Esaki, p. 644]
1959	Integrated Circuit (IC)	[Sah, p. 1290]
1962	Laser Diode	[Harman, p. 363]
1962	Light-Emitting Diode (LED)	[Dummer, p. 166]
1967	Semiconductor Memory	[Sah, p. 1301]
1971	Microprocessor	[Rotsky, pp. 21–22]
1978	Optoelectronic IC (OEIC)	[Lee *et al.*, p. 806]

Only the tunnel diode, discovered by Esaki of Japan, breaks the long U.S. line of important device innovations. Based upon a concept described in 1954 by Shockley, a coinventor of the transistor [Shockley, pp. 799–826], the tunnel diode is the first of modern microwave semiconductor devices. It is ironic that the only important semiconductor device originating from Japan never achieved success in the marketplace. The tunnel diode was a technical milestone only. Because it has limited power output and frequency range, subsequent microwave device innovations from the U.S. eclipsed the tunnel diode.

Three of the eight innovations are optoelectronic semiconductors: the laser diode, the light-emitting diode (LED), and the optoelectronic IC (OEIC). The leading edge semiconductor technology in the merging of computers and communications is the optoelectronic semiconductor.

Optoelectronics

Optoelectronics is the branch of electronics dealing with the portion of the electromagnetic spectrum that encompasses infrared, visible, and ultraviolet light. An optoelectronic semiconductor is a device that detects, amplifies, and transmits light.

Until the mid-1980's, the importance of optoelectronic semiconductors was not understood by most writers of semiconductor policy. In general, Western analysts and policymakers seemed unable to grasp the notion that photons, elementary particles of light, will share with electrons the task of transmitting and processing information in the 21st century. The authors of a book on telecommunications described in 1977 the major product innovations in the semiconductor industry to 1968, but failed to mention either the laser diode or the LED [Sciberras and Payne, p. 53]. In 1980, a book examining innovation, competition, and government policy in the semiconductor industry did not mention optoelectronic semiconductors, illustrating only a family tree of semiconductor technology whose branch denoting discrete devices is divided into "transistors," "diodes," and "others" [Wilson et al., p. 20]. An otherwise well-researched book in 1985 [Malerba, pp. 13–14] about the semiconductor business graphically divides semiconductors "into three major product groups: discrete devices, optoelectronic devices, and integrated circuits." In contrast to the other two product groups, the optoelectronic device classification contains no device subcategories. No mention is made of the laser diode, the most important of the optoelectronic semiconductors, or the photodiode, the device used for detection. An intelligent analysis of the semiconductor industry in the U.S.

and Japan, written in 1984, notes that "if gallium arsenide came to be viewed as superior to silicon in optoelectronics," then incremental increases in the Japanese lead in random-access memories (RAMs) would not be as significant as "this revolutionary new technology" [Weinstein *et al.* p. 69].[1]

This is an astute observation, but it should have been couched in imperative rather than problematical terms. Elemental silicon to date has not been an efficient optoelectronics material when compared to gallium arsenide (GaAs). GaAs was the material used in the invention of all three optoelectronic device innovations—the laser diode, the LED, and the OEIC.

The Marketplace

In contrast to American observers, Japanese analysts understood early the importance of optoelectronics. In 1977, a Professor from the University of Tokyo noted that the outstanding innovations in information technology include optical fibers and laser diodes [Ishii, p. 84]. In 1979, the government sponsored the establishment of the first national optoelectronic device R&D project, designed to bring into commercial production the optoelectronic devices conceived and developed by U.S. innovators.

The failure of Japanese firms to create innovative devices has not stopped them from wresting market leadership from the U.S. in all of the above-mentioned U.S. discoveries, except the microprocessor and OEIC. The U.S. leads in microprocessor shipments because the Japanese strengths in product development and manufacturing technology are less important for microprocessor applications than are the American strengths in architectural design and software development. The OEIC is not yet in volume production.

[1]By periodically introducing denser RAMs in high volumes priced to sell at anticipated maximum yields, Japanese semiconductor firms have captured the electronic memory market. These denser RAMs result from incremental technological improvements in an established device structure. The authors believe that gallium arsenide (GaAs) optoelectronics represents a revolutionary new technology. They did not understand that the superiority of GaAs in optoelectronic applications already had been established.

The Japanese, usually reluctant to document the U.S. loss in international competitiveness, have revealed some alarming figures concerning the decline of U.S. production. The Electronics Industries Association of Japan (EIAJ) reported that U.S. electronic production had decreased from a 50.4% worldwide share in 1984 to 39.7% in 1987, while Japan increased its share from 21.3 to 27.1% ["Electronics industry finds unlikely support of claims," p. A6]. During the same period, the U.S. share of worldwide semiconductor shipments decreased from 44.7 to 31.5%, while Japan's share went from 21.3 to 39.1% [Instat Inc. and Semiconductor Industry Association, p. 29]. In 1986, Japan replaced the U.S. as the world leader in optoelectronic components, increasing its market share from 39% in 1981 to 62% in 1986, while the U.S. share dropped from 53 to 30% [Nagasawa and Forrest, p. 24]. It is no coincidence that these declines parallel each other.

Future U.S. Competitiveness

This book examines the U.S. competitive position in OEICs and makes specific recommendations that will help restore U.S. semiconductor competitiveness, an essential condition if the U.S. is to obtain its technological leadership in the 21st century. The focus is on the OEIC because it represents the next milestone in the technological development of microelectronics for the communication and computer industries [Wada, pp. 471–474]. Most examinations of U.S. and Japanese semiconductor policy are concerned with technologies already well established in the marketplace. U.S. policymakers are asked by self-serving interest groups to spend pounds of money after market leadership has been lost. It is more cost-effective to spend ounces of money on emerging technologies that can supplement or complement existing technologies. Semiconductor competitiveness is essential if the U.S. is to maintain its economic preeminence in the information age. The semiconductor represents both a critical enabling technology and a core industry. As the leading democratic power in the world, the U.S. cannot become dependent upon foreign sources in key technologies and industries. The U.S. must have the technological and economic freedom necessary to preserve its national security and well-being in this uncertain world of countries and ethnic groups burdened with long histories of hate and strife.

Because OEICs are not available commercially, the U.S. still has an opportunity to become the market leader. Answers to the following questions cast light on what the U.S. can do to compete in the marketplace:

1. How far ahead of the U.S. is Japan in OEIC development?
2. What gives Japan its advantage in OEIC development?
3. What are the U.S. deficiencies in OEIC development?
4. What Japanese or new policies should the U.S. adopt?

To answer these questions, I shall describe OEIC technology and the market it serves, review the activities of individual American and Japanese firms in OEIC research and development (R&D), examine the comparative positions of the U.S. and Japan in OEIC R&D expenditures, analyze the technical output on OEICs by both countries in leading scientific and engineering publications, and describe American and Japanese optoelectronic policies. Finally, I shall recommend courses of action that the U.S. can take.

U.S. government and industry policymakers will find this book useful in formulating and implementing strategies to restore U.S. competitiveness in semiconductors and in optoelectronic devices. Industry observers and analysts will understand why the OEIC is an important development. I shall demonstrate that the OEIC, which combines optical and electronic functions on the same chip, is an important advance in the progress of microelectronics, the 20th-century technology most responsible for changing the social and economic face of the world. "The success of lightwave communications depends critically on the development of a viable monolithic optoelectronic integrated circuit (OEIC) technology" [Ketterson *et al.*, pp. 73–76].

2

Optoelectronic Semiconductors

Technology

Semiconductors are discrete, monolithic, or hybrid devices. A discrete semiconductor is a distinct, unconnected device that contains a single active element, either a transistor or a diode, and performs only one function, such as emitting light (laser diode) or detecting light (photodiode). The finished product is a unitary package or single structure. A monolithic device is an integrated circuit (IC), capable of performing two or more different functions. It contains two or more active elements integrated on a single chip, and fabricated at the same time. A hybrid device is a single structure containing two or more interconnected discrete or monolithic chips, fabricated individually and mounted on a common substrate. Semiconductor technological progress generally is measured from the time it takes to proceed from a discrete device to a monolithic IC. The latter is considered superior to the other two structures in most applications, and represents an advance in the state of the art. For instance, the performance level of a system using discrete laser diodes driven by electronic ICs is improved by integrating on a chip the laser diode and the electronic circuits that drive it [Nakamura *et al.*, pp. 822–826]. The head of Photonic Device Research at Bellcore, the research arm of the regional Bell operating companies, states that "opto-electronic integrations promise to have higher performance, more functionality, better reliability, and lower cost than discrete devices" [Lee, p. 269].

Because a semiconductor has optical properties, it is used in a number of optoelectronic modes—laser diode, light-emitting diode (LED), or photodetector. A laser diode is an electroluminescent device, i.e., it converts electrical energy to light energy. It emits a coherent, highly directional, and intense beam of visible or infrared (IR) light. The dominant emission is not spontaneous, but rather stimulated when the injection current (electrons) exceeds the threshold current level at which point laser action occurs. Hence, we have the acronym *laser*, for Light Amplification by the Stimulated Emission of Radiation. The laser power output is proportional to the drive current above the threshold. A LED also is electroluminescent, emitting visible or IR light that is spontaneous and incoherent. The LED has no threshold current level, emitting light that is almost linearly proportional to the drive current.

Photodetectors include photodiodes, phototransistors, photoconductors, and photovoltaic cells. Light falling on the junction area of a photodiode will generate an electrical current if it is connected to a circuit. A phototransistor is simply a photodiode with an additional junction, giving it an amplifying property. A photoconductor is a bulk effect device, having no junction area. It increases in electrical conductivity as the light level increases, and decreases in conductivity as the light level decreases. A photovoltaic cell generates a voltage across its junction when light falls on it. A well-known example is the solar cell. Fiber optic systems employ laser diodes, LEDs, and photodiodes.

A monolithic OEIC integrates optoelectronic devices and electronic circuits on a single semiconductor substrate. A photodiode integrated with an amplifying circuit is a photoreceiver. If a laser diode is integrated with a photodiode to monitor its power output and/or an electronic circuit to drive it, a photoemitter is the result. If these functions are contained in the same package, but fabricated individually and then interconnected on a common substrate, a hybrid OEIC is the result.

The material properties of a semiconductor define to a great extent the basic characteristics of the device. Optoelectronic semiconductors are fabricated from a variety of materials, the most important being the III–V intermetallic compound, gallium arsenide (GaAs). Elemental silicon has been the dominant electronic semiconductor material, but its device range in optoelectronics has been limited to use as a solar cell or as a photodetector. It has not been used as a laser diode or LED. The movement of electrons in silicon causes energy losses primarily in the form of heat (phonons) rather than light (photons). These energy losses make it inefficient to use silicon devices as light emitters. In GaAs, radiation emission is in the form of photons. A layer of GaAs, 1 micron (μm) in thickness, may absorb up to 80% of the available photons. It would require a silicon layer about 100 times as thick to absorb the same proportion of photons. However, scientists have not given up on silicon. By varying its chemical composition, researchers at the University of Rochester have demonstrated optical emission in silicon at very low efficiencies [Bradfield *et al.*, pp. 100–102]. Defense researchers in the United Kingdom have announced luminescence in silicon by altering its physical structure [Jackson, p. 25, June 1991]. In both cases, emission was stimulated by optical rather than electrical means. Electroluminescence was obtained by researchers at the Fraunhofer Institute for Solid State Technology in Munich [Richter *et al.*, pp. 691–692]. These early research results give no indication of the efficacy of silicon as a viable laser material. Device efficiency is too low for practical use, and process and device stability have not been established.

Another important property of GaAs is its capability of being converted into ternary or quaternary compounds by the addition of other elements during processing. This results in devices with different and more versatile properties. The primary reason in optoelectronics for adding elements to GaAs is to vary the wavelength of operation. For instance, the peak emission wavelength for GaAs at room temperature is 870 nanometers (nm). The wavelength is not useful in fiber optic applications because it is not an optimal transmission point for optical fibers. By adding aluminum (Al) to form AlGaAs on a substrate of GaAs, the emission wavelength can be shifted to 820 nm, a point where fiber losses for short-haul transmissions such as computer networking can be held to a minimum.

Indium phosphide (InP) is the material used for long-haul transmissions, and is the material of choice in telephony. InGaAs or InGaAsP on an InP substrate operates at the peak emission wavelengths of 1300 or 1550 nm, optimal transmission windows for optical fibers, where attenuation losses and scattering effects are held to a minimum. Although both emitter and receiver OEICs in InP have been demonstrated in the laboratory [Suzuki *et al.*, pp. 1479–1487], InP to date has not been as versatile a semiconducting material as silicon and GaAs. Aside from its use in laser diodes and photodiodes, the only InP semiconductor presently sold commercially is a discrete bulk-effect microwave device, the Gunn diode. There are no InP logic devices as yet. InP transistors have exhibited higher speeds than have silicon and GaAs devices ["Hughes makes fastest InP transistor yet," p. 28], but cannot be fabricated structurally in the same way as standard GaAs transistors because of basic material properties. InP semiconductors are more tolerant of radiation and demonstrate less degradation in space applications than do GaAs and silicon devices [Messick, p. 31]. The Department of Defense (DOD) funds much of the effort to make InP a more versatile semiconductor.

Scope of Market

At the present time, most optoelectronic semiconductors sold commercially are discrete devices. Some hybrid OEICs are available, consisting of small numbers of discrete devices on a nonsemiconducting substrate. Matsushita did introduce a transmitter chip in 1989, integrating a laser diode and an IC driver circuit of three transistors, and a receiver chip containing a four-transistor IC amplifier [Shibata and Kajiwara, pp. 34–38]. Both were designed for fiber optic applications in cable television and intrabuilding telephone communications. However, these OEICs apparently are used only for internal applications because they still cannot be purchased commercially. Antel Optronics, a small Canadian supplier, introduced late in 1989 a two-function monolithic silicon OEIC receiver, consisting of a photodiode integrated with a preamplifier circuit for fiber optic local area networks (LANs) and backplane connectors ["Integrated FO receiver," p. 178]. However, no GaAs OEICs have been introduced.

The optoelectronic semiconductor market in this book will include the following:

Laser diodes

LEDs

Photodiodes

Hybrid OEICs

Phototransistors

Photoconductors

Photovoltaic cells

Optocouplers

Imaging arrays

Integrated optics

An optocoupler, also known as an optoisolator, usually consists of a device module containing an LED and a photodiode or phototransistor, with the emitter and detector electrically insulated from each other, but optically coupled through an electrical insulator such as air, plastic, or an optical fiber. Its purpose is to isolate elements of an electrical circuit. Often included in this market category are optoelectronic sensors. They have the same basic configurations as optocouplers, but are used for such applications as object presence or motion sensing. An array is a single substrate containing a number of discrete or integrated devices, such as photodiodes, all performing the same function. Included in this market category are certain imaging devices that do perform more than one function, such as the sensing of light and the processing and storage of electronic signals.

The category of integrated optics consists of hybrid passive and control devices whose operation is totally optical. They include waveguides, switches, modulators, directional couplers, filters, and lenses. Although development is being conducted on the use of GaAs in integrated optics [Boyd, pp. 743–846], there are no commercially available monolithic optical logic devices. The ultimate goal is to develop an optical integrated circuit in which information is encoded into light for processing. Optics is the third leg of the computing triad of electronics, optoelectronics, and optics. The importance of optoelectronic and optical computing lies in their capability to execute tasks impossible for electronic computing to accomplish. The inherent parallelism of optics will allow the construction of highly parallel processors in which large amounts of information can be processed simultaneously. Optical computing may play a future role in the development of a wide variety of advanced applications such as artificial intelligence and neural networks, and other attempts to model the human brain. Most researchers and analysts in the U.S. and Japan agree that optical information processing is in the early research stage [Ikegami and Kawaguchi, pp. 1131–1140; West, pp. 34–46; Midwinter, p. 10; Bell, pp. 34–57; Silvernail, pp. 127–129; Katauskas, pp. 32–36]. Many in the optoelectronics community question the capability of optical data processing to replace electronic data processing. The weak interaction characteristic which makes optics so desirable for information transmission works against it for information processing. Although GaAs is the most logical choice for the first generation of optical computers, researchers still disagree on the definition of optical computing, what materials will be used, and what the architecture of the devices and systems will be. The concept of optical computing is more exotic, and receives more exposure than the OEIC in the popular media [Hooper and Schlesinger, p. 1], but the OEIC is the next important device advance in information technology.

Size of Market

According to Japan's Optoelectronics Industry and Technology Development Association (OITDA), Japanese industry produced

TABLE 2-1 Worldwide Optoelectronic
Semiconductor Market by Geographical Area
(1988)

	$ Millions	% Share
Japan	1751	70
U.S.	505	20
Europe	246	10
Total	2502	100

Source: Author's estimate based on industry
sources and on Kales, p. 73, June 1989.

$1751 million worth of optoelectronic semiconductors in fiscal
year (FY) 1988 [Kales, p. 73, June 1989]. Based on the figures
in Table 2-1, in 1988, Japan produced 70% of the world's optoelectronic semiconductors, compared to 20% for the U.S.

In 1976, the U.S. produced 68% of the world's optoelectronic semiconductors, compared to 10% for Japan. By 1981,
when the first Japanese optoelectronics project became operational, Japan's share had increased to 39%, while the U.S. share
had declined to 53%. In 1986, Japan surpassed the U.S. with a
worldwide share of 62%, compared to 30% for the U.S. [Na-
gasawa and Forrest, p. 24].[2] Figure 2.1 illustrates the decline in
market share of the U.S. and Europe.

The Japanese semiconductor industry achieved the com-
manding position in the optoelectronic device market by concentrating on LEDs in the 1970s, and laser diodes in the 1980s.
It was assured of a major share of the LED market because of
its dominance in consumer goods and its victory in the calcu-
lator wars. Laser diode production has been driven by the demand for digital audio disks, compact disks, video disks, copiers,
facsimile, and printers. Two-thirds of the $15 billion Japanese
output in 1988 of optoelectronic components, equipment, and
systems is in the equipment category. Optical disk drives and
input/output equipment account for 83% of this category or

[2]OITDA reports its optoelectronics figures by application rather than
devices. It includes optoelectronic semiconductors in a component category
with display devices, connectors, and lasers other than laser diodes. However,
figures for optoelectronic semiconductors can be identified by careful analysis.

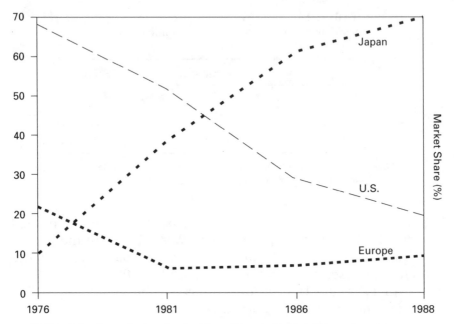

Fig. 2.1 Optoelectronic Semiconductor Market Shares by
Geographical Area (1976–1988).

Source: Author compilation based on figures from OITDA and
industry sources.

$8.35 billion, and communications equipment for only 5%. Although laser diodes represented 28% of the world laser market in 1986, they accounted for 82% of the Japanese market. The key device structure in OEIC development is the laser diode.

The Japanese semiconductor firms aggressively pursue optoelectronic equipment markets, pulling along device production. The economies of scale that are realized in device production increase the value added in equipment production. Japanese firms that market both optoelectronic devices and equipment gain a larger share of the device market by concentrating on high-volume production and shifting profits from the device level to the equipment or system level.

In contrast to the Japanese market, at least 60% of U.S. laser diode production is for the communications market. U.S. laser diode production is characterized by low volume, expensive packaging, and high unit prices. There is no U.S. consumer industry to offer experience in high-volume device production in preparation for the demand for computer OEICs. Because the typical U.S. optoelectronic semiconductor firm is not part of a large horizontally integrated organization, there is no internal market for which large-scale production can be initiated, nor is there a corporate bank upon which to draw funds for the substantial capital expenditures required for high-volume OEIC production.

Japanese optoelectronic device manufacturers fund capital expenditures with less risk because of in-house demand. OEIC R&D expenditures are based upon strategic corporate goals rather than individual market performances. Short-term market cycles are ignored. The Japanese semiconductor strategy of building for long-term growth means that they obtain lower prices and better deliveries for semiconductor manufacturing equipment during periods of semiconductor recession. They apparently have learned from the great American entrepreneur, Andrew Carnegie, that the best time to build for a market up-swing is before it starts [Heilbroner and Singer, p. 160].

The Importance of the OEIC

Industry observers have grasped the concept that fiber optic transmission is superior in speed and capacity to conventional electrical transmission. Optical signals travel longer distances without repeaters. A higher frequency range and bandwidth provides signal capacities unmatched by electrical transmission. A single fiber cable containing two fibers can handle 1300 two-way conversations, compared to 24 for two copper twisted pairs. Optical cable is immune to radio interference and free from grounding or shorting problems. If a fiber cable breaks, there is no electrical shock, sparks, or fire hazard. Fiber cable is thinner than twisted pair, and weighs less than half of a twisted pair per 1000 ft. Security breaches are detected easily because data must be physically removed from optical cable, decreasing signals and increasing error rates. All of this is easily understood by policy writers and makers because the concepts deal at a system level with a familiar technology: telephony.

When an industry observer looks into the internal workings

of a computer, concepts are not understood as readily. Even within the industry, it is often difficult for an engineer or manager to rise above his technological paradigm or immediate self-interest. Silicon device engineers tend to ignore the potential of GaAs ICs and optical interconnects. Management is busy protecting its silicon markets and investments, trying to wring out the last bit of performance from a mature technology at a sometimes prohibitive cost both in money and time.

The electronic IC basically requires only the additional integration of laser diodes and photodiodes to become functional as an OEIC. The OEIC addresses major performance problems in electronically based computers, including the most important, interconnection delays. The continuing miniaturization of computers because of increasing chip densities and circuit complexities has strained the capacity of electronic interconnects in the system hierarchy. From machine to module to board to chip, and even on chip, the use of optical interconnects will reduce system delays and improve reliability [Tsang, p. 122, 1987].

Within computer networks, it is the interconnections that are increasing costs and degrading performance. Inexpensive desktop computers can process data at speeds of 640 million bits per second (640 Mb/s). Ethernet, the most popular data communications network, typically operates at 3 Mb/s, with a limit of 10 Mb/s. The national defense network was upgraded partially in 1989 to 1.5 Mb/s. It has been proposed to operate a national network at 45 Mb/s. The proposed Fiber Distributed Data Interface (FDDI) standard will operate at 100 Mb/s.

These speeds are inadequate for interfacing with today's computers. The Intel 80386 32-bit microprocessor operates at 33 million operations per second (MOPS). This is an effective speed of 1 billion gigabits per second (1 Gb/s). To take advantage of this speed, optical interconnects must be implemented between computers, and within the computer. There is no advantage to using high-speed microprocessors and central processing units (CPUs) if this processing power becomes lost at the local bus, backplane, and peripheral bus connections.

The performance of a high-speed computer chip depends to a great extent on how fast the signal can be moved off the chip. Designers have tried to compensate for off-chip signal delays by increasing the speeds of transistors on the chip, and by increasing the amount of transistors per chip. The scaling down of the transistors does increase speeds and densities, but also has increased the size of the chip, adversely affecting yields and reliability.

Not all interconnections become shorter with the scaling down of feature dimensions. Smaller transistors on larger chips complicate circuit layout and routing. The distances from transistors to the edge of the chip, where the connections from chip to chip reside, have tended to become longer. In addition, an increase in transistor densities reduces transistor power limits, unless the total power that the chip can dissipate is increased. However, higher chip dissipations incur increased signal delays. With the increased use of speedy GaAs ICs, interconnection delays become even more critical. There is a fine line of compromise which the designer must straddle to optimize performance.

The problem is understood better when one considers that a chip consists of three distinct parts: active elements, conductors, and insulators. The active element (transistor) is where the change in conductivity occurs upon the application of an external voltage. The conductors can be considered as passive wiring (interconnects). The passive insulators isolate the active elements. In a 16-kb CMOS static random-access memory (SRAM), 80% of the cell area is devoted to interconnects, 14% to insulation, and 6% to active elements. If relative volumes are considered, the interconnect region occupies 67%, insulation 31.5%, and the active region 1.5% of the chip [McGreivy, p. 211]. The key to increasing system throughput is to concentrate on reducing interconnect delays. For many applications, it is not cost-effective to spend time and money trying to increase the speed of the active element.

If on-chip signals were not routed to the edge of the chip, but were removed immediately after processing, circuit design complexity and signal delays would be reduced materially. At the Seventh International Conference on Integrated Optics and Optical Fiber Communication, Kobe, Japan, in July 1989, AT&T Bell Laboratories and Bell Communications Research (Bellcore) jointly reported the development of an array of two million surface-emitting GaAs laser diodes fabricated with a semiconductor lithographic printing technique on an area of 1 cm^2 [Carts, pp. 23–26]. Researchers have known since the mid-1970s that for OEICs, surface-emitting laser diodes are preferable to the edge-emitting devices now used in discrete and hybrid applications. Surface emitters can be accessed more easily, and a variety of transmission media can be employed. Surface-emitting laser diodes first were reported in 1965 by M.I.T. Lincoln Laboratory, and were demonstrated as a possible structure for OEICs in 1974 by Bell Laboratories. However, progress in the integration of laser diodes with electronic ICs proceeded slowly because of the differences in structure and composition between the laser diode and the field-effect transistor (FET), a construction first used in the fabrication of GaAs ICs. Laser diodes for communication applications are vertical structures fabricated on a conductive substrate, while the FET is a horizontal structure on a semi-insulating substrate. In addition to the lack of structural compatibility and the early problems associated with integrating different semiconductor materials, laser diodes did not have the small size and low power dissipation needed for high-density devices. This new development by Bell Labs/Bellcore can overcome many of the barriers associated with the integration of electronic and photonic functions on a single chip.

Although Bell Northern Research (BNR), a Canadian firm, developed at its R&D facilities in Research Triangle Park, NC, the first OEIC transmitter using semiconductor techniques to fabricate edge-emitting laser diodes in 1988, the power dissipation of the laser diodes was too high and the sizes too large for use in the high-density chips needed in today's computer systems [Bindra, pp. 43–44]. The key in the Bell Labs/Bellcore development is the simultaneous downsizing of the individual laser diodes for decreased power dissipation, and the use of standard photolithographic processing techniques to fabricate the array. The two million diode array can be packed onto an area about the size of a fingernail. The array was fabricated by the same epitaxial process used by BNR, molecular beam epitaxy (MBE), a technique for depositing ultrathin layers (50 Å) of differing materials in making advanced GaAs ICs.

Bell Labs/Bellcore expect that further development will

result in the lowering of the current threshold of the individual laser diodes in the array so that as many as 200 million laser diodes can be packed together in this fingernail-sized chip. Because of power dissipation limits with the present array, only about 20 laser diodes can be operated continuously at one time.

Most companies engaged in the development of surface-emitting laser diode arrays have had telecommunications or military applications in mind. The objective is to produce arrays with improved brightness per unit area by increasing power output without increasing the size of the emitting aperture, i.e., the hole in the surface through which the light is transmitted. This can be achieved by increasing diode efficiency or decreasing thermal resistivity. A number of U.S. and Japanese firms have been working on high-density arrays, but none seems suitable for data processing. A leader in this area has been the David Sarnoff Research Center, Princeton, NJ. This former unit of RCA, now a subsidiary of SRI International, Menlo Park, CA, has developed a two-dimensional (2-D) monolithic grating surface-emitting (GSE) laser diode array that can be integrated with electronic functions by means of an uninterrupted quantum-well waveguide which couples the photonics to the electronics [Evans *et al.*, pp. 97–106]. Rockwell has demonstrated an OEIC transceiver that outputs light through a similar grating [Hutcheson *et al.*, pp. 34–35]. However, this structure does not appear to be as effective as that of Bell Labs/Bellcore for integration with high-speed computer chips.

The development by Bell Labs/Bellcore may prove to be a milestone in OEIC development. By using standard IC wafer-processing techniques to integrate laser diodes on an electronic IC, the circuit designer can route signals to the nearest laser diode rather than to the edge of the chip. If chips can be optically connected through free space, then more functions can be crammed on the chip.

Bell Labs and Bellcore are research laboratories devoted to communications applications. GaAs OEICs have their greatest application in intracomputer connections and computer networking. Bell Labs is a unit of AT&T. Bellcore is the research arm of the regional Bell operating companies. It is only natural that they are more interested in advancing communications technology. Both are spending a great deal of effort on indium phosphide (InP) OEICs which are better suited for information transmission and reception than for information processing. InP is in its early stage of development as a versatile semiconductor material. No InP integrated circuits (ICs) have been made outside of the laboratories. InP currently is about ten years behind GaAs as an IC material.

Typical of the development work performed by Bellcore is the recent demonstration of an InP OEIC with a transmission rate of 5 Gb/s over a distance of 29 km [Lo et al., pp. 673–674]. This is an advance for multimedia communications, but not for electronic data processing. However, AT&T may shift its development sights with the acquisition in 1991 of NCR, an important computer manufacturer.

Wafer-scale integration (WSI), the use of the uncut wafer to create an IC, may become a viable alternative to the diced and packaged chips of today. For over 25 years, engineers and scientists have been trying to reduce interconnect densities by connecting the chips on the wafer, before they are diced into individual chips [McDonald et al., pp. 32–39]. WSI eliminates the dicing step.

Several semiconductor and computer companies through the years have made attempts to develop WSI devices for the commercial market, using electrical interconnects, but with no success. The latest attempt is being made by a small British firm, in cooperation with Fujitsu, a leading Japanese semiconductor supplier. Fujitsu supplies the processed wafer and packaging. The British firm provides the electrical interconnect software and associated logic (Anamartic). Research and development on WSI is being conducted by a number of universities and firms interested in parallel-processing computers [Chin, p. 20]. Three-dimensional (3-D) circuits are being investigated in which several wafer-scale circuits are stacked and interconnected optically. This packaging technique reduces the size of the computer and increases its speed.

Early in 1991, IBM announced a WSI development, the batch fabrication of laser diodes on a 2-in.-diameter GaAs wafer using standard integrated circuit manufacturing techniques [Jackson, p. 32, May 1991; Troy, p. 32]. Researchers at IBM's laboratory in Zurich were able to etch laser mirrors into the wafer through a batch-processing technique, rather than individually forming the mirrors on each minute laser diode after the wafer is diced into individual devices. In addition to the 5000 laser diodes, plus the photodiodes and test devices now on the wafer, the researchers believe that they can integrate other electronic and optical elements.

The increased use of GaAs and high-speed silicon ICs for faster processing exacerbates the problem of off-chip interconnect signal delays. Because of the speedy rise times of these ICs, the off-chip connection lengths at which said connections must be treated as active transmission lines rather than as passive wiring become much shorter. Circuit design becomes more complex when an interconnect must be treated as an active rather than a passive element. There is no advantage to using faster devices if off-chip transmission line delays become greater than device delays. A typical rise time for a GaAs IC is 100 picoseconds (ps). This means that the maximum interconnection length for passive wire design is 0.30 in. Beyond that length, the designer must consider the interconnect as an active transmission line. The fastest silicon IC technology, ECL (emitter-coupled logic), has a typical rise time of 600 ps, indicating a critical interconnect length of 1.80 in. Older and slower silicon IC designs, such as TTL (transistor–transistor logic), exhibit rise times on the order of 8000 ps, allowing for maximum interconnection lengths of 24.0 in. [Pound, p. 51].

The problems relating to the use of higher speed ICs do not decrease as the signal moves further from the chip. The various operating units of a computer are connected by printed circuit boards and copper wire. Impedance noise is created by signal reflections from mismatched impedances and ground noise. Ground noise increases with the use of faster logic and more backplane connections and transmission lines. Crosstalk becomes more noticeable with the decrease of rise times [Markstein, pp. 48–50]. Circuit designers now must treat printed circuit boards and passive wiring as active elements just like the ICs which they connect. As computers become reduced in size, these problems are multiplied. Optical interconnects offer the circuit designer an opportunity to pack data channels much closer without incurring the increased throughput delays and system noise that would result if electrical interconnects were used.

IBM has demonstrated the most complex OEIC to date. Designers from the Zurich Research Division and the Thomas J. Watson Research Center in Yorktown Heights, NY, reported a GaAs transceiver chip set consisting of an OEIC receiver and a hybrid transmitter in 1988 [Hecht, pp. 22–26]. The receiver contains an 8000-transistor electronic IC integrated with an array of four photodiodes. The hybrid transmitter consists of two chips, an array of four laser diodes, and a 5000-transistor IC. The very large scale integrated (VLSI) transceiver performs a variety of functions including timing, multiplexing, and demultiplexing.

The IBM OEIC is designed for computer systems that need to exchange complex images and data in billions and trillions of operations per second (BOPs and teraops). Computer applications require higher density OEICs than do telecommunications applications because the key operation is data processing rather than signal amplification. Computer links operate over shorter distances where higher integration is more important than receiver sensitivity. This IBM development is useful for linking board-level and chip-level circuits within the computer, and increasing the speed and reliability of the computer operating units. All types of computers will benefit, from supercomputers and mainframes to workstations and personal computers.

The IBM transceiver operates at 1.0 Gb/s over the 830–850 nm spectral range. The receiver chip is about 1/4-in. square with minimum feature dimensions of one micrometer. Byte-wide data streams are combined by the hybrid transmitter into a serial stream for optical fiber transmission by the laser array. The optical signals are converted by the receiver's photodiode array into electronic signals, and then separated, reformed, and retimed by the electronic elements on the receiver chip. The four optical fibers are sliced at an angle and positioned directly above the photodiodes for maximum efficiency. The electronics on the receiver chip include a preamplifier, a decision circuit, a clock-recovery circuit, and a 1:10 demultiplexer. All of these functions now are marketed as separate ICs by AT&T and Japanese companies. Although IBM is not a merchant[3] semiconductor supplier, it is preserving its position as the world's largest computer manufacturer by an accelerated program to develop VLSI OEICs for internal use.

[3]The U.S. semiconductor market is divided into two categories: merchant and captive. The merchant market is served by semiconductor producers whose primary goal is to supply other companies rather than in-house requirements. The captive semiconductor market comprises those producers whose manufacturing facilities are dedicated to supplying devices for corporate applications. The U.S. has a large captive semiconductor market, estimated to be around 30% of the total amount of semiconductors produced by U.S. firms. Captive suppliers include IBM, thought to be the largest semiconductor manufacturer in the world, General Motors, Digital Equipment Corporation, and many other computer, communications, and defense firms.

3

U.S. and Japanese Semiconductor Firms Developing OEICs

History

The concept of integrating optical components on a single substrate first was advanced by a Bell Telephone scientist in 1969 [Miller, pp. 2059–2069, 1969]. The OEIC first was proposed by a researcher from the California Institute of Technology (Cal Tech) in 1973 [Yariv, p. 1656]. A Cal Tech group reported the development of the first OEIC in 1978 [Lee *et al.*, pp. 806–807]. Figure 3.1 lists milestones in the development of OEICs. The U.S. dominated early optoelectronic device and OEIC R&D. In spite of this, none of the leading U.S. merchant semiconductor firms, except AT&T, currently has a recognizable, commercially oriented OEIC development program at this writing. In contrast, all of the leading Japanese semiconductor firms are developing OEICs.

1969	First conception of the monolithic integration of optical components	AT&T
1973	OEIC first proposed	Cal Tech
1978	First recognized OEIC	Cal Tech
1979	First OEIC optical repeater	Cal Tech
1980	First demonstrated OEIC receiver	AT&T
1985	First reporting of a multichannel OEIC	Fujitsu

Fig. 3.1   History of the Development of the OEIC.
Sources: Miller, pp. 2059–2069, 1969; Yariv, p. 1656; Lee *et al.*, pp. 806–807; Yust *et al.*, pp. 795–797; Leheny *et al.*, pp. 353–355; Makiuchi *et al.*, pp. 634–635.

U.S. and Japanese Developers

U.S. semiconductor manufacturers, not including small firms, but including IBM, a captive producer only, identified as developing OEICs are:

AT&T
Harris
Honeywell
Hughes
IBM
Motorola
Rockwell
TI
TRW
Westinghouse

All of these companies, except AT&T and IBM, receive a substantial amount of their funding from the DOD.[4] This funding

---

[4]Technical papers on OEICs by these companies often acknowledge partial funding by agencies, commands, or laboratories of the DOD, such as Honeywell does in a 1986 paper wherein it is noted that "this work is partially funded by the Defense Research Projects Agency and U.S. Army Strategic Defense Command" [Ray *et al.*, p. 104]. Other companies such as Rockwell refer in the text of the paper to the needs of military and space applications

is not part of a concerted or specific effort to develop a manufacturing capability for OEIC devices, but is incidental to programs with their own missions.

Harris, Motorola, and TI are major U.S. semiconductor suppliers with no present commitment to manufacture OEICs. The government sectors of these companies usually conduct the R&D on OEICs. In 1991, Harris sold the optoelectronic semiconductor operation that it acquired from General Electric, who had acquired it from RCA. Honeywell [1986] has been reducing its merchant semiconductor and electrooptic activities ["Loral agrees to buy defense subsidiary from Honeywell Inc.," p. B14]. IBM does not produce for the commercial market. Hughes, TRW, and Westinghouse are second-line semiconductor producers, more concerned with supporting their defense operations. TRW sold its optoelectronics subsidiary in 1988.

The unfortunate fact is that the major U.S. semiconductor suppliers, with the exception of AT&T, are not committed to the development of OEICs. Many do not manufacture optoelectronic devices or GaAs ICs. Figure 3.2 lists the top ten U.S. merchant suppliers of semiconductors in 1988 and the optoelectronic devices which they sell. AT&T is listed as tenth because only about 25% of its total semiconductor sales were to external customers [Shandle, pp. 125–126]. If captive sales were included, then it would rank sixth behind Harris. Not only do just three of the top ten manufacture optoelectronic devices, but only AT&T produces laser diodes and GaAS photodiodes, the key device structures in the development of OEICs.

The largest U.S. optoelectronic semiconductor supplier, Hewlett Packard, currently has no OEIC development program. The company's fourth-place ranking in worldwide optoelectronic semiconductor sales in 1988 was not sufficient to propel it into the top U.S. semiconductor ranks. Hewlett Packard derives its strength in optoelectronics from its leadership in LED technology.

Figure 3.3 lists the ten leading Japanese semiconductor

[Kilcoyne *et al.*, p. 148] or make reference to DOD sponsorship [p. 153], and cite DOD contracts [p. 155]. Many of the papers are written by personnel from the government sectors or corporate research centers of the corporation. Few come from the semiconductor sector.

	Laser Diodes	LEDs	Photodiodes
Motorola		X	X*
Texas Instruments		X	X*
Intel			
National			
AMD			
Harris			
LSI Logic			
Signetics			
Analog Devices			
AT&T	X	X	X

* Silicon only.

Fig. 3.2 Optoelectronic Devices of the Top Ten U.S. Merchant Semiconductor Suppliers in 1988.

Sources: Compiled by author from data in: Instat Inc., p. 30, Sept. 25, 1989; Instat Inc., p. 20, Oct. 16, 1989; Dataquest Inc., p. 31, Aug. 1990; company annual reports and sales literature.

suppliers, all of whom are performing OEIC R&D. These companies all produce optoelectronic devices. Figure 3.3 shows only the optoelectronic devices that the companies are aggressively selling into the U.S. market. The U.S. clearly is lagging behind Japan because it does not have a broad base of major semiconductor firms developing OEICs and selling optoelectronic devices.

Optoelectronic semiconductor technology is being developed intensely by the Japanese, who believe that it is critical to their plan for domination of the world's information industries [Tassey, p. 2]. By the end of the first decade of the 21st century, practically all long-distance and most local signal transmission will be optoelectronically based. More importantly, the OEIC will be one of the standard solutions to the problems of increasing the throughput and reliability of computer systems. It is difficult to understand why the major U.S. semiconductor suppliers have not placed more emphasis on the development of OEICs. Some of them are introducing silicon ICs for lightwave circuits, but without laser diode capability, they have little chance to compete against Japanese firms, who can offer a com-

	Laser Diodes	LEDs	Photodiodes
Nippon Electric (NEC)	X	X	X
Toshiba	X	X	X
Hitachi			
Fujitsu	X	X	X
Mitsubishi	X	X	X
Sanyo			
Sharp	X	X	
Oki	X		
Sony	X		

Fig. 3.3 Optoelectronic Devices Sold in the U.S. by the Top Ten Japanese Semiconductor Suppliers in 1988.

Sources: Compiled by author from data in: Instat Inc., p. 25, Sept. 25, 1989; Dataquest Inc., p. 31, Aug. 1990; company annual reports and sales literature.

plete solution to an application. The small American firms are the leaders in the fight to regain U.S. optoelectronics leadership.

Status of Small U.S. Firms in Optoelectronics

Figure 3.4 lists the small firms in the U.S. that are producing laser diodes. Also noted is their status as LED and photodiode suppliers. Ortel, Epitaxx, and Spectra Diode are the most outstanding technically, and have been the recipients of DOD contracts. Ortel is a spinoff from Cal Tech where the OEIC first was conceived and demonstrated. The President of the company was a member of the group that developed the first OEIC. PCO Inc. originally was a joint venture of Plessey, a British supplier of defense, consumer, and semiconductor products, and Corning Glass, the pioneer developer of optical fibers. Plessey then sold its interest to Corning. In 1988, IBM acquired a 25% share, leaving 65% for Corning, and reducing the share of the founders and employees from 20 to 10%. In 1991, PCO announced that it was ceasing operations by the end of the year. Lytel was acquired in 1989 by AMP Inc., a major U.S. producer of both standard and fiber optic connectors. Laser Diode, formerly an

	LEDs	Photodiodes
Epitaxx	X	X
Fermionics		
Laser Diode	X	
LaserCom		X
Lasertron		X
Lytel	X	X
Ortel		X
PCO	X	
Spectra Diode		

Fig. 3.4   Small Laser Diode Firms in the U.S.
Sources: Company sales literature.

operating unit of M/A-COM Inc., a leading U.S. supplier of microwave communications systems and modules, was sold in 1986 to Morgan Electronics, a U.S. subsidiary of Morgan Crucible of the U.K. In addition to the above, there are U.S./foreign ventures selling laser diodes and associated products in the U.S. market. They include Amperex, an American semiconductor supplier purchased many years ago by the Dutch giant, Philips, and BT&D, a joint venture of Dupont and British Telecom, the latter a major telecommunications producer and an OEIC developer.

These small U.S. firms have capable technical and management personnel, but do not have the capital to fund OEIC development and production. They can package discrete devices as optical receivers and links, but they must purchase the electronic ICs. They have limited product lines, and are trying to stay afloat in a market dominated by AT&T and the large Japanese firms. There are a number of other small American suppliers of LEDs and photodiodes, but they have little possibility of being a factor in the OEIC market without laser diode manufacturing technology.

4

Optoelectronic Semiconductor R&D Expenditures in Japan and the U.S.

Definitions

Research and development (R&D) expenditures are used by analysts as a statistical tool to determine technological advances. R&D is a term whose meaning can differ from industry to industry, and from firm to firm. Detailed breakdowns among industry, academia, and government rarely are available. Analysts do not clearly define what is research and development [Flamm, p. 191]. Semiconductor firms rarely categorize R&D according to product.

I shall use the classification of research and development that has been advanced by the Organization for Economic Co-operation and Development (OECD). It delineates three categories of R&D [Dumbleton, p. 8].

Basic research: original investigation for new scientific knowledge, not specifically directed.

Applied research: original investigation for new scientific knowledge, specifically directed.

Experimental development: using existing scientific knowledge to produce new products or substantial improvements in existing ones.

In the semiconductor industry, experimental development includes product design, product engineering, product development, and process and/or manufacturing engineering. Semiconductor firms differ on how they account for process or manufacturing engineering costs. Some or all of these costs are allotted to the cost of manufacture. Companies may define process engineering costs as those associated with wafer fabrication (front end) and include these costs under development. Assembly, test, and package (back end) may be allocated to the cost of manufacture. The front end is considered the more technologically sophisticated area. Front-end engineers earn more than back-end ones. Research engineers and scientists are paid higher salaries than product developers. These differences tend to skew comparisons in semiconductor R&D expenditures, but generally not to an unacceptable extent. Absolute figures may vary according to the source, but trends usually are discernible.

Trends in R&D

The amount of money spent by individual companies in the U.S. and Japan on R&D for optoelectronic devices or OEICs is not available. Tassey of the National Bureau of Standards, in his Planning Report 23, dated October 1985, has estimated the 1981–1987 optoelectronic R&D expenditures of the U.S., Japan, and Europe (see Appendix, Table A-1). He arrived at his figures by asking industry and government sources "to estimate the number of full-time scientists and engineers engaged in optoelectronic R&D for firms actively pursuing this technology," and by reviewing MITI documents [29–30]. Tassey's figures include optical fibers, GaAs ICs, and passive connectors and couplers [7]. These products are not optoelectronic semiconductors and represent separate markets. Optoelectronic R&D expenditures of companies such as Dupont, Kodak, and Polaroid are included in Tassey's estimate.

While the actual figures are not useful in determining optoelectronic semiconductor or OEIC R&D expenditures, Tassey's estimate does indicate that Japan had an earlier commitment to optoelectronics than the U.S. Using Table A-1, I have illustrated graphically in Fig. 4.1 that Japan's spending level was higher than that of the U.S. until 1987. In 1981, when Japan's first national optoelectronics development project, the Optoelectronics Joint Research Laboratory (OJRL), became operational, it spent 60% more than the U.S. on optoelectronics R&D. By 1986, when the second national optoelectronics project, the Optoelectronics Technology Research Laboratory (OTRL), was announced by Japan, U.S. spending virtually had reached Japan's level. With the ability to overcome the "NIH" (not invented here) factor, the Japanese built early on American research by devoting most of their effort to the development of

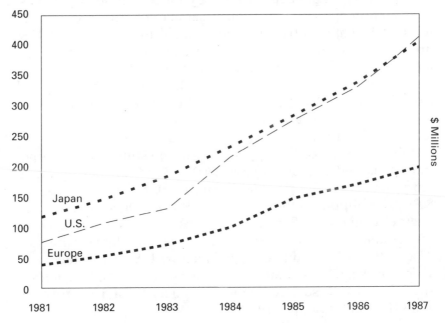

Fig. 4.1 Worldwide Trends in Optoelectronic R&D Expenditures.

Source: Tassey, p. 30, Table 2, Oct. 1985.

GaAS and InP devices [32]. The U.S. stepped up its spending in response to the Japanese national optoelectronic projects. Expenditures were increased by 82% from 1983 to 1984 as the results of the first project came in. After the second Japanese optoelectronics project started, the U.S. increased R&D spending by 74% from 1986 to 1987.

Industry R&D

Aggregate indicators by themselves do not offer useful data for a particular technology [Miller, p. 201, 1982]. Table A-2 in the Appendix shows the corporate R&D expenditures of the five leading merchant semiconductor producers in both the U.S. and Japan in 1988. The total sales of these corporations divided by their corporate R&D expenditures indicate that the U.S. firms spent 9.4% of their sales on R&D, a figure nearly 50% higher than the 6.4% expended on R&D by the Japanese companies. High-technology products such as semiconductors are R&D intensive. A significantly larger amount of the sales of the U.S. firms are in semiconductors, when compared to those of the large horizontally integrated Japanese firms. The percentage of semiconductor sales to total sales by the Japanese firms ranged from 7 to 19%. For the U.S. firms, the range was 33 to 100%. In this case, corporate R&D as a percentage of sales is a misleading indicator for semiconductor R&D expenditures.

In Table A-3 of the Appendix, I have estimated the corporate-sponsored semiconductor R&D expenditures of the same top producers in the U.S. and Japan. The U.S. firms spent about 80% ($1.6 billion) of their aggregate R&D budgets on semiconductor technology. Semiconductor R&D as a percentage of semiconductor sales in 1988 for these leading firms, who sold 82% of the semiconductors produced by U.S. companies, was 14.7%. This figure is about 50% higher than the median of 9.7% spent by all U.S. semiconductor suppliers from 1976 through 1986 [Howell *et al.*, p. 219].

It is more difficult to determine the semiconductor R&D expenditures of the top Japanese producers because no company reported semiconductor sales larger than 19% of total sales. I have estimated in Table A-3 that the top five Japanese firms, whose sales represented 92% of the total Japanese output in 1988, spent about $2.7 billion on semiconductor R&D or 16.4% of their semiconductor sales. This figure is slightly higher than the median of 15% expended from 1973 through 1974 [Kimura, p. 64].

There is not a significant amount of difference between the percentage of semiconductor R&D to semiconductor sales for the top firms in the U.S. and Japan in 1988. Does this mean that optoelectronic semiconductor R&D expenditures as a percentage of semiconductor sales for the top U.S. and Japanese suppliers were roughly equal? It does not because three of the five U.S. firms do not sell optoelectronic devices.

Total semiconductor R&D expenditures for the five Japanese firms were $2.715 billion, 70% more than their U.S. counterparts. These firms were able to spend more on semiconductor R&D in 1988 because they made greater profits. Corporate profits of the five Japanese companies were $13.770 billion during the five-year period from 1984 through 1988, compared to $3.094 billion for the five U.S. semiconductor leaders ["Looking at the leaders," pp. 30–72]. In addition, year-to-year profits of the vertically and horizontally integrated Japanese companies were steadier and more predictable than the top U.S. semiconductor suppliers. During this five-year period, no Japanese company had a loss year, while three of the five U.S. firms suffered an aggregate of six loss years. National and AMD had five of the six. The Japanese firms had more R&D funds available, but where they were spent is unknown. It may be that the percentage of sales expended on semiconductor R&D shows commitment, but the Japanese ability to spend more may be a better indicator of results. The Japanese showed higher commitment than the Americans in the 1970s and early 1980s, when the median R&D percentage for all Japanese semiconductor producers was about 50% higher than the median for U.S. firms.

Corporate R&D spending is more closely related to the existing or potential market to which it is directed, and how well the corporation is doing in the marketplace. In this instance, the market figures for the top five semiconductor firms of each country are not reliable indicators, even if they were available, because all of the Japanese firms are optoelectronic device suppliers, whereas Motorola and Texas Instruments are the only U.S. suppliers. Intel, National, and AMD produce no optoelectronic devices, and have no discernible OEIC R&D program. Therefore, I shall use the optoelectronic semiconductor market figures of the U.S. and Japan in Table 2-1 in Chapter 2 to estimate optoelectronic R&D expenditures.

U.S. semiconductor R&D as a percentage of corporate revenue in 1988 was 9.4% [Rayner and Stallman, p. 58]. Multiplying that figure by U.S. optoelectronic semiconductor sales of $505 million, I estimate that corporate-sponsored expenditures for optoelectronic semiconductor R&D in 1988 for U.S. industry was $47 million. Because Japanese sales were 3.5 times that of the U.S., I estimate that Japan firms spent $165 million on optoelectronic semiconductor R&D in 1988.

It is clear from the figures that the top Japanese semiconductor producers have a greater commitment to and capability for optoelectronic semiconductor R&D than do their U.S. competitors. If the U.S. cannot depend upon its leading semiconductor suppliers to compete against Japan in optoelectronics, can it depend upon its OEIC developers? Table A-4 in the Appendix compares the corporate-sponsored R&D expenditures of the top Japanese OEIC developers with the leading U.S. developers, with the exception of IBM, which does not participate in the merchant semiconductor market, and Ortel, which is too small to be of significance.[5] It again illustrates that Japanese semiconductor firms have more money for funding the commercial development of OEICs.

AT&T is the only U.S. firm with the capital and commitment to challenge the Japanese giants. TRW, Westinghouse, Rockwell, and Honeywell are major defense contractors, which accounts for their low percentage of corporate-sponsored R&D expenditures. TRW ranked seventh in fiscal year (FY) 1988 in defense electronic sales, Westinghouse ranked 11th, Rockwell

---

[5]In 1990, a billion dollar Japanese firm invested $11 million in Ortel for a 21.6% equity share and the right to distribute Ortel products in the Far East ["Ortel signs agreement with Sumitomo Cement," p. 26]. Sumitomo Cement is diversifying into optoelectronics and electronics. Ortel will use the funds from Sumitomo to increase production in CATV laser diode devices and modules. Why did not a U.S. semiconductor producer invest in this developer of the first OEIC?

was 12th, and Honeywell 25th [Callan, p. 63]. Rockwell and Honeywell have received significant GaAs IC funding from the DOD, but have not been able to convert their device development and pilot production lines into commercial successes. All of the above-mentioned defense contractors, plus TI, which ranked 13th, had defense electronic sales of at least $1.01 billion in the U.S. government fiscal year 1988. Motorola, ranking 40th, had defense electronic sales of about $500 million. The figures in Table A-4 cannot be used as a basis for estimating optoelectronic semiconductor R&D expenditures, but they do indicate that the Japanese firms have "deeper pockets" from which to make choices about commercial OEIC development.

Table A-4 also indicates the deep Japanese commitment to optoelectronics. The five top Japanese semiconductor producers, plus the sixth-ranking Matsushita, are aggressive suppliers of optoelectronic semiconductors. The addition of Matsushita, the world's tenth largest semiconductor supplier in 1988, to the Japanese list serves only to strengthen the probability that Japan will be the early leader in the OEIC market. In addition, Sanyo, Oki, and Sharp are three other Japanese optoelectronic suppliers that represent formidable competition. They are members of the ten-year project to develop OEICs. Sanyo ranked 14th in world semiconductor sales, and Oki was 15th. Sharp advertises itself as the number one optoelectronics supplier in the world ["We are the block," pp. 58–59].

The bottom line is that there is no commitment at this time on the part of the U.S. semiconductor producers, except for AT&T, to produce OEICs for the merchant market. The Japanese semiconductor producers show not only commitment, but also R&D capability and merchant marketing expertise, all backed up by greater capital resources and a cooperative government. Seven of the top ten Japanese semiconductor firms are in the world's leading 20 electronics companies, and four of them rank in the first five [Kaplan, p. 80]. The only U.S. company in this group is Motorola, ranking 19th.

Government R&D Funding

Except for AT&T, all of the U.S. semiconductor suppliers in Table A-4 performing OEIC R&D received most of their funding from the DOD and Department of Energy (DOE). IBM, a captive supplier, also spent a significant amount of corporate funds on OEIC R&D. Total federal funding for semiconductor R&D in FY 1987 was $454 million [Congressional Budget Office, pp. 59–72]. Table 4-1 shows that $29 million or about 6.5% of this funding was for optoelectronic devices. Over half of the funding was from the DOE for photovoltaic material and devices. Material funding included GaAs, InP, and GaAs on a silicon substrate (GaAs/Si). Some of this work will be useful in the development of OEICs. The next largest category was mercury cadmium telluride (HgCdTe) imaging or sensing arrays. This is a priority item for high-performance sensing applications such as the Strategic Defense Initiative (SDI) where it is used for the detection of the optical radiation resulting from exhaust plumes during the boosting phase of an enemy ballistic missile. HgCdTe devices offer little commercial promise in the foreseeable future, except for selected applications in thermal imaging and night vision where cost is not a factor.

It is not possible to trace the amount that the U.S. government funded annually on R&D for OEIC technology during FY 1987 and FY 1988. Spending for R&D on OEICs is buried in a number of programs, many of which are classified. Even governmental agencies are not able to obtain precise spending figures from other agencies of the federal government. The DOD's second report to Congress on critical technologies estimates that its funding for photonics from FY 1986 through FY 1990 by its Science and Technology Program in the DOD and DOE amounted to $560 million. However, a footnote cautions that the funding figure "is of the right magnitude, but is not to be

TABLE 4-1 U.S. Government Funding of Optoelectronic Semiconductor R&D in FY 1987

	$ Millions	%
Department of Defense (DOD)	12.8	43.7
Department of Energy (DOE)	15.0	51.2
Other Agencies	1.5	5.2
Total	29.3	100.0

Source: Author's compilation based on: Congressional Budget Office, pp. 59–72.

construed as a precise budgetary quantity" [Department of Defense, p. A-70, Mar. 15, 1990].

Japanese government funding for optoelectronic semiconductor R&D in FY 1987 and FY 1988 is estimated at $5.7 million annually, based upon a ten-year budget of $82 million for the second national optoelectronics project, of which 70% will be provided by the government's Key Technology Center [Sakurai, p. 104]. Most of this money will be spent for OEIC development.

Comparison of R&D Expenditures in the U.S. and Japan

Table 4-2 estimates total R&D expenditures for optoelectronic semiconductor technology in Japan and the U.S. for 1988. U.S. government figures are for the 1987 fiscal year which starts October 1, 1987 and ends September 30, 1988. The Japanese corporate fiscal year ends in March 1988. The figures indicate that Japan is spending over twice as much as the U.S. on optoelectronic semiconductor R&D. The U.S. government spends five times that of the Japanese government, but it is in industry where the spending is more effective in transforming technology into a marketable product.

TABLE 4-2 Optoelectronic Semiconductor R&D Expenditures in the U.S. and Japan in 1988

	$ Millions
U.S.	
Industry	47.0
Government	29.3
Total	76.3
Japan	
Industry	165.0
Government	5.7
Total	170.7

Source: Author's estimate.

5

Scientific Writing

The data analysis so far indicates that Japan is leading in the development of the OEIC. Japan dominates the market for discrete optoelectronic semiconductors (see Table 2-1). Preeminence in the sales of discrete semiconductors historically has not guaranteed leadership in ICs in the United States. Except for Texas Instruments, none of the leading U.S. discrete semiconductor manufacturers in the 1950s now is a leading IC supplier. The reverse is true in Japan where all but one of the top six discrete device suppliers in 1959 continue to be the leading IC suppliers. These same Japanese firms now are the leaders in discrete optoelectronic semiconductor devices. There is no reason to believe that they cannot continue their success with OEICs.

In Chapter 3, I have shown that all of the leading Japanese semiconductor manufacturers are engaged in the development of OEICs, whereas in the U.S., only AT&T and IBM, the latter a captive producer, have serious development programs for the commercial market. Military requirements drive OEIC development at the other large U.S. firms. Small U.S. firms are not capitalized sufficiently to be a market threat.

Japan is spending more than twice as much as the U.S. on optoelectronic R&D, and 3.5 times as much in the commercial area (see Table 4-2). Because the Japanese consider the OEIC as "one of the most exciting topics in the field of semiconductor optoelectronic device research" [Wada et al., p. 805], it is safe to speculate that a significant amount of Japanese funding is being directed towards OEIC R&D.

As another indicator for determining if the Japanese are leading in OEIC development, I have examined papers on OEICs from five publications during the years 1983–1988. These five were selected because they are the leading English-language publications for timely papers on experimental development in optoelectronic device technology. The year 1983 was chosen as the starting point because the only monthly publication in the U.S. devoted entirely to lightwave technology began in that year. In addition, the period 1983–1988 includes the realization by U.S. researchers of the importance of the OEIC as a result of the development activity arising from Japan's first national optoelectronics project.

There is evidence that the number of technical papers published by a country can be linked to its technological progress [Narin and Frame, pp. 600–605]. No indicators bearing directly on OEIC technology were available to me. Development expenditures for OEICs are not in the public domain. The numbers of personnel engaged in OEIC development are not known. Patent data for OEICs have not been tabulated.

Aggregate indicators are available, but do not offer useful data for a particular technology [Miller, p. 20, 1982]. From 1965 to 1983, the U.S. had more scientists and engineers than any other country, except the U.S.S.R. [National Science Board, p. 5]. However, from 1979 to 1983, the Japanese organized and conducted most of the R&D that brought them to market leadership in optoelectronics in the early 1980s.

Aggregate patent activity from 1969 to 1982 indicates that the U.S. led in external patent applications [National Science Board, p. 9], although Japan rapidly was closing the gap. Because optoelectronics is a fast-moving technology, many small U.S. companies operate with the philosophy of *take the tech-*

nology and run. They perceive that patent activity is a tool of protectionism for the Japanese government, and a sales tool for Japanese manufacturers.

Four of the five selected publications are engineering journals concerned with product development. Although the fifth is published by a scientific society, it also is a good source for optoelectronic product developments. I included publications directed towards device development rather than basic and applied research. Although Japan is clearly behind the U.S. in both of the latter areas [Narin and Frame, p. 604], it has not affected Japan's ability to become the undisputed leader in the technology and marketing of optoelectronics devices. Three of the five selected journals are published by the Institute of Electrical and Electronics Engineers.

The Institute of Electrical and Electronics Engineers, Inc. (IEEE)

The IEEE is the largest engineering society in the world, with over 300 000 members, including 60 000 who reside outside the U.S. Members of the IEEE annually attend more than 5000 worldwide IEEE professional meetings, conferences, and conventions. According to the *1990 IEEE Annual Report* [pp. 6A–6D, May/June 1991], the IEEE published 24% of the world's technical literature on electrical engineering technology in 1989. Most of these papers are of U.S. origin. In Table 5-1, I have selected eight from the total of 19 technologies listed by the IEEE to show that there is a relationship between papers published and technological and/or market position.

Japan is the world leader in the manufacture and sale of radio, television, and audio equipment. Table 5-1 indicates that the IEEE percentage of papers published in this industry is significantly lower than the average percentage for all papers in electrical engineering. Printed and hybrid circuits and passive circuit components, also dominated by the Japanese, are below the average. The only other technology under the average is electro-optics and optoelectronics, the lowest.

The highest percentage of IEEE papers were on magnetic

TABLE 5-1 Percentage of the World's Technical Literature in Electrical Engineering Published by IEEE (December 31, 1989)

	Percent
Printed and Hybrid Circuits	15.1
Radios, Television, and Audio	14.6
Electro-Optics and Optoelectronics	12.0
Passive Circuit Components	23.0
Semiconductor Materials and Devices	25.9
Microwave Technology	36.1
Information and Communication Theory	37.1
Magnetic and Superconducting Materials and Devices	37.9
Average of All IEEE Electrical Engineering Literature (19 Topics)	24.0

Source: 1990 IEEE Annual Report: The Year in Review, p. 6B.

superconducting materials and devices, information and communication theory, and microwave technology. The first field is still in its conceptual stage, the period when the U.S. typically is dominant. The U.S. is the world leader in the second highest field, information and communication theory, and is dominant in microwave technology because of the U.S. military programs in radar, electronic warfare, and electronic countermeasures. The percentage of papers published in semiconductor materials and devices indicates a middle ground, consistent with the U.S. technological and market positions.

Because the percentage of IEEE papers published on electro-optics and optoelectronics falls below the average for all topics, and Japan leads in the markets corresponding to these topics, these data tend to validate Japan's claim that it is the world leader in optoelectronic components. In addition, the percentage of IEEE papers published on electrooptics and optoelectronics declined from 17.5% in 1987 to 12.0% in 1989. During the same period, the average of all electrical engineering literature published by the IEEE dropped from 28 to 24% [*1988 IEEE Annual Report*, p. 6B]. It should be noted that the percentage of IEEE papers published is not the same as the percentage of U.S. papers. Although the overwhelming majority of IEEE papers are written by Americans, authors from Japan, Europe, and the rest of the world are represented.

Publications Analyzed

The following five professional publications were selected for analysis.

Journal of Lightwave Technology

Articles on current research and development, applications, and methods used in lightwave technology and fiber optics. Topics include optical fibers, photoemitters and photodetectors, waveguides, integrated optics, hybrid circuits, and systems and subsystems employing these active and passive devices. This journal is the premier professional journal in English on optical and optoelectronic topics. Articles are lengthy, using numerous illustrations and citations. Most articles are about optical fibers, laser diodes, and integrated optics. Published jointly by the IEEE and the Optical Society of America (OSA). Publication started in January 1983. Issued quarterly in 1983, bimonthly in 1984 and 1985, and monthly beginning in 1986.

IEEE Journal of Quantum Electronics

Articles on quantum electronics and applications, including optoelectronic theory and techniques, lasers and fiber optics, and on systems and subsystems design, development, and manufacture employing quantum and optoelectronic techniques. This publication includes electronic semiconductor topics, and concentrates primarily on laser diodes in the optoelectronic device area. Published monthly by the IEEE.

IEEE Electron Device Letters

Rapid publication of two-to-three-page articles on the theory, design, development, performance, and reliability of electron devices, including tubes, semiconductors, energy sources, power devices, and displays. Most of its articles describe advanced electronic integrated circuit technology. Published monthly by the IEEE.

Electronics Letters

A British publication similar to *IEEE Electron Device Letters*, but with more emphasis on optical and optoelectronic topics. Its two-page articles describe technologies found in the other three journals. It has the widest technological coverage, and is published 25 times annually by the Institution of Electrical Engineers (IEE).

Applied Physics Letters

Rapid publication of new developments in general physics and its application to other sciences, engineering, and industry. Published semimonthly by the American Institute of Physics (AIP) with the cooperation of the American Physical Society (APS) and OSA.

Quantitative Analysis

Table 5-2 is a breakdown of the papers on OEICs by geographical area published in the five selected English language technical journals from 1983 through 1988. Although this book is concerned only with the U.S. and Japan, Europe and the rest of the world (ROW) are included for comparison. Of the 72 papers

TABLE 5-2   OEIC Papers by Geographical Area

	1983	1984	1985	1986	1987	1988	Total
U.S.	2	3	5	5	2	7	24
Japan	5	5	8	12	5	4	39
Europe	1	0	0	4	2	1	8
ROW*	0	0	0	1**	0	0	1
Total	8	8	13	22	9	12	72

     *Rest of World.
   **Canada.

Sources: Author's compilation from 1983–1988 issues of: *Journal of Lightwave Technology* (IEEE); *IEEE Journal of Quantum Electronics*; *IEEE Electron Device Letters*; *Electron Letters* (IEE); *Applied Physics Letters* (AIP).

published, 54% originated from Japan, 34% came from the U.S., 11% from Europe, and 1% from the ROW.

Table 5-2 indicates that 1986 was a peak year for OEIC papers. This corresponds to the ending of the first MITI-sponsored optoelectronics R&D project, which began in 1979 and ended in 1986. Of the total output during the six-year period, 30% were published in 1986. Figure 5.1 graphically illustrates that the Japanese output peaked in 1985–1986, during which 51% of all Japanese papers were published. Personnel from six of the seven leading Japanese semiconductor companies in the first national optoelectronics project published OEIC papers in the five journals. Japan probably will publish another flurry of papers at the end of the second optoelectronic project in the 1995–1996 period.

Table 5-3 shows the origin of the papers within each geographical area. Industry accounts for 58 of the 72 papers, a percentage share of 81%. Japanese firms hold a 66% share of the OEIC papers published by industry worldwide, whereas U.S. industry has a 31% share. This table is the best quantitative

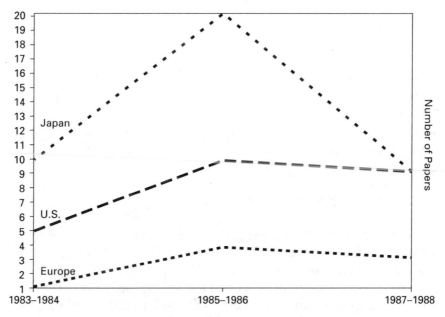

Fig. 5.1 Worldwide Trends in the Publication of OEIC Papers.
Source: Author's compilation from same sources as in Table 5-2.

TABLE 5-3   OEIC Papers by Industry and University/Government

	1983	1984	1985	1986	1987	1988	Total
U.S.							
AT&T	0	0	4	0	1	4	9
Rockwell	0	2	0	0	0	0	2
IBM	0	0	0	1	0	1	2
Motorola	0	0	1	1	0	0	2
Honeywell	2	0	0	0	0	0	2
Ortel	0	1	0	0	0	0	1
Industry Total							18
Univ./Govt.	0	0	0	3	1	2	6
U.S. Total							24
Japan							
Fujitsu	4	2	7	6	1	1	21
NEC	0	2	0	2	2	1	7
Hitachi	1	0	0	2	1	0	4
Toshiba	0	0	0	1	0	1	2
Mitsubishi	0	0	0	1	1	0	2
Matsushita	0	1	0	0	0	0	1
Sumitomo	0	0	0	0	0	1	1
Total Firms							38
Univ./Govt.	0	0	1	0	0	0	1
Japan Total							39
Europe							
Br. Telecom	0	0	0	1	0	0	1
Thomson–CSF	0	0	0	1	0	0	1
Univ./Govt.	1	0	0	2	2	1	6
Europe Total							8
ROW							
Univ./Govt.	0	0	0	1	0	0	1
World Total	8	8	13	22	9	12	72

Source: Author's compilation from same journals as in Table 5-2.

indicator of the relative progress of U.S. and Japanese semi-conductor firms in the development of OEIC technology. Corporate R&D is closer to the marketplace than that of universities or government laboratories. Six of the seven Japanese firms are major semiconductor producers. NEC and Toshiba ranked first and second, respectively, in worldwide merchant semiconductor sales in 1988 [Instat Inc., p. 30, Sept. 25, 1989], Hitachi was third, Fujitsu sixth, Mitsubishi eighth, and Matsushita tenth. Sumitomo Electric, the only firm not producing semiconductor devices for the merchant market, is a major supplier of optical fibers and optical data links, and the world's leading producer of semiconductor material. On the U.S. side, Motorola ranked fifth in semiconductor sales. IBM is not a merchant producer, although its internal sales volume probably puts it at number one worldwide. Rockwell and Honeywell primarily are suppliers for military systems, with at least half of their output going to captive applications. AT&T is trying to become a leading merchant semiconductor supplier, but while its total volume placed it in 12th place in 1988, only 25% of that was sold in the merchant market. Ortel, founded by researchers from the California Institute of Technology, has excellent R&D credentials, but is a very small company, funded in part by contracts from the DOD.

In the university/government category in Table 5-3, one of the six U.S. university/government papers comes from the Naval Research Laboratory, the leading government semiconductor laboratory. The Japanese paper is from a university. Four of the six European noncorporate papers are from a French national telecommunications laboratory conducting R&D under ESPRIT, the European Economic Community (EEC) program for technological competitiveness. The other two come from universities in England and Germany. The ROW paper is of Canadian origin.

Table 5-3 shows that Fujitsu leads in the number of papers published on OEICs, AT&T is second, followed closely by NEC, the world's largest merchant semiconductor producer. Fujitsu has published 55% of the 38 papers originating from Japanese industry, and 36% of the industry total of 58 papers worldwide. AT&T ranks second, representing one half of the U.S. industrial output, and 16% of industry worldwide. NEC has authored 18% of the total papers from Japanese industry, and 12% of industry worldwide. Japan has published 66% of the total papers from industry, and 54% of the papers from all worldwide sources.

Qualitative Analysis

Table 5-4 classifies U.S. and Japanese industry papers by content. All papers describe OEICs for communications applica-

TABLE 5-4 Analysis of the Content of Industry OEIC Papers (U.S. and Japan)

	Transmitter	Receiver	Optical Repeater	Switch Connect	Review
AT&T	3	4	0	1	1
Rockwell	2	0	0	0	0
IBM	0	2	0	0	0
Motorola	0	1	0	1	0
Honeywell	0	2	0	0	0
Ortel	0	0	1	0	0
U.S. Total	5	9	1	2	1
Fujitsu	5	12	2	1	1
NEC	4	1	1	0	1
Hitachi	2	0	0	0	2
Toshiba	1	0	0	0	1
Mitsubishi	1	1	0	0	0
Matsushita	1	0	0	0	0
Sumitomo	0	1	0	0	0
Japan Total	14	15	3	1	5
Total	19	24	4	3	6

Source: Author's compilation from same journals as in Fig. 5.1.

tions, except for one by Motorola in the switch/connect category that is designed for signal processing applications. Although there is little in the literature about OEIC technology for computer systems, it is this field that ultimately offers the largest market for OEICs. However, the first OEICs will be for fiber optic communications applications.

Motorola designed a coupling and packaging technique using a silicon photodetector in which the input is optical and the output is electrical [Hartman *et al.*, p. 73, Jan. 1986]. The other two devices in the switch/connect category include a 12-channel individually addressable detector and a LED emitter interconnect array by AT&T [Ota *et al.*, pp. 1118–1122], and a four-channel switch module by Fujitsu [Iwama *et al.*, pp. 772–778]. The Fujitsu switch contains a four-channel OEIC receiver, a 4 × 4 GaAs IC electronic switch, and a four-channel OEIC laser diode transmitter. It is the most practical of the OEIC devices in this category.

Twelve of Fujitsu's 21 papers are on photoreceivers, the

OEIC that will be first in volume production for communications applications because it immediately offers a higher level of integration than the phototransmitter. Four of AT&T's papers are on photoreceivers. Both companies, along with NEC and Ortel, have published papers on optical repeaters. Optical repeaters represent an advancement in OEIC technology because they integrate both phototransmitters and photoreceivers on the same substrate.

Conclusions

Based upon the analyses of the papers in Tables 5-3 and 5-4, Fujitsu is the furthest advanced in OEIC technology for deployment in communications systems. These figures correlate with the opinions of industry sources and my readings of other professional journals and periodicals concerned with OEICs. They indicate that AT&T is the leader in applied research, and shares the leadership in experimental development with Fujitsu in optoelectronic device technology. AT&T has a slight lead in product design, but Fujitsu is ahead in the development of marketable products, and is leading in manufacturing technology. NEC also is strong in manufacturing technology, and is an excellent developer of a wide variety of semiconductor products.

Honeywell has been a leader in OEIC development for communications systems, describing in 1983 a monolithic phototransmitter that integrated a laser diode with its driving circuit and a 4:1 multiplexer [Carney et al., pp. 48–51]. However, much of Honeywell's R&D is funded by the DOD, which discourages the publication of technology deemed sensitive for national defense. Rockwell researchers, also funded by the military, authored a paper in 1983 that describes an OEIC transmitter integrating a laser diode with a two-transistor driving circuit [Kim et al., pp. 44–47]. Neither of the companies has authored papers on OEICs since 1984 in the five publications analyzed. From time to time, news of their activities does appear in journals and periodicals. A Rockwell transceiver was reported in 1987 [Hutcheson et al., pp. 34–35]. Light is emitted by vertically constructed laser diodes through a grating similar to that developed by the Sarnoff Center, previously described in Chapter 3. The same article reported that Honeywell was developing a 1000-transistor transceiver chip.

Fujitsu is an aggressive semiconductor supplier with strength in communications and data processing. AT&T is strong in communications and data processing, but it is weak in semiconductor marketing, with most of its semiconductor production going to captive applications. Honeywell has been reducing its effort in the merchant semiconductor market. Ortel is a pioneer in OEIC R&D, but is too small a company to mount a large-scale production and marketing effort without reorganization and external assistance. Motorola has an optoelectronic product line focused on LEDs and optocouplers. The firm is not committed yet to OEICs. Both of its papers were authored by its Government Electronics Group, which implies that the technology is being directed towards captive military applications.

Except for AT&T, there is no serious U.S. challenger to Japan at this time in OEICs for the communications market. All of the Japanese semiconductor producers will perform creditably, with NEC probably fighting Fujitsu for the top spot. NEC is not only the world's largest merchant semiconductor producer, but is a major supplier of millimeter/microwave semiconductors to the Japanese communications market and the U.S. communications and military markets.

There is insufficient data to determine which merchant semiconductor supplier leads in OEICs for the information processing market. Fujitsu, Japan's largest computer manufacturer, and third largest in the world in 1988 [Kelly, p. 28], and now the second largest with its purchase of a leading British computer firm, certainly will use its know-how to develop and market computer OEICs. AT&T was ranked tenth in the U.S. and 21st worldwide in information processing in 1988 [Kelly, p. 22], and has a long history of introducing important electronic and semiconductor innovations. Its ability to successfully compete against the Japanese in the computer market still is in question. IBM probably is the technological leader for computer OEICs because of its domination of the computer industry and its excellent reputation in IC manufacturing technology. However, IBM is a captive producer only. It influences the merchant semiconductor market primarily as a customer because it purchases some of its requirements from outside vendors.

6

Japanese Optoelectronic Policies

Introduction

All of the policies described in this chapter are concerned with the acquisition, dissemination, and protection of technologies that contribute to Japanese success in the optoelectronic device marketplace. It is not within the scope of this study to examine every policy of the Japanese government and industry that has influenced the course of optoelectronics. I have selected those policies that have been particularly successful. An important factor in this success has been the active cooperation between government and industry. The conception and implementation of policies on semiconductor technology in Japan are based upon a long planning stage during which all parties strive to reach a consensus for action. They are not a hasty reaction to perceived crises. Japan uses a "bottom-up" approach in its decision-making process to develop a national strategy. Japanese industry and government cooperate closely in targeting technologies and performing R&D. Fierce competition begins after the technology is transferred to the individual firms.

Targeting and Protecting

The only major U.S. merchant semiconductor firm seriously committed at this time to the commercial development of OEICs is AT&T, whereas all of the leading Japanese companies are committed. How did this happen?

A 1982 report by the Office of the U.S. Trade Representative identifies *targeting* as a Japanese government industrial policy "formulated in support of a national strategy of designating particular industries as essential for the development of the Japanese economy" [Office of the U.S. Trade Representative, pp. 48–53]. The report identifies the Ministry of International Trade and Industry (MITI) as the Japanese government agency having the lead responsibility for targeting. Japanese government officials state that it is inaccurate to pose the idea of a targeted industry [Okimoto, p. 97n]. Instead, high-priority technologies, which cut across industrial boundaries, are selected for targeting.

What is the meaning of the term *technology*? The dictionary defines it as "industrial science; systematic knowledge of the industrial arts." A more precise definition is a "body of knowledge about the production of goods and services" [Barke, p. 7]. High-technology industries, such as the semiconductor industry, are defined as "those characterized by large research and development (R&D) expenditures and rapid technological progress" [Nelson, p. 1].

The Japanese perception of how they target is correct. U.S. government policymakers fail to note the importance of the Agency of Industrial Science and Technology (AIST), an extra-ministerial office of MITI. The AIST was established in 1948 to oversee technology in industry and mining. Its responsibilities in 1962 were extended to include the making of science and technology policy [Mirzoeff, p. 41]. Many of the Japanese technical articles referenced in this book were sponsored by the AIST, which was responsible for the research and development activities of Japan's first national optoelectronics project [Hayashi *et al.*, p. 1432].

A review of U.S. literature reveals that many analysts fail to distinguish between technology and the industrial sector. A very few understand the mission of the AIST. A Professor of International Business at Sophia University in Tokyo, coauthor of a well-known book on the Japanese corporation and an astute observer of Japanese management styles, failed to distinguish between technology and the industrial sector. He notes that electronics is the "sector that the Japanese government is most determined to support and defend," but that except for petroleum, it is the sector most penetrated by foreign investment [Abegglen and Stalk, p. 225]. In fact, it is the technology and not the industries that Japan is protecting. IBM and Texas Instruments (TI) are major players in the Japanese electronics market only because they refused to sell their vital computer and semiconductor technologies for cash only. The only way that the Japanese could license their technologies was to allow them to establish wholly owned manufacturing facilities in Japan. Japan would have been excluded from exporting ICs to the U.S. if it could not obtain patent rights from TI [United States International Trade Commission, p. 60]. IBM and TI obtained licenses for 100%-owned facilities in spite of the fact that it was a period when the law prohibited foreigners from owning more than 49% of any computer or IC venture in Japan [Flamm, p. 254].

Other U.S. companies with innovative technologies that can be circumvented have not enjoyed the success of TI and IBM in establishing production facilities in Japan. A U.S. power transistor manufacturer established a joint venture in the mid-1950s to sell its products and those of the joint venture in Japan and Southeast Asia. The joint company, whose stock ownership was divided evenly between the U.S. and Japanese partners, had difficulty in obtaining the favorable domestic financing arrangements available to wholly owned Japanese firms. In 1974, the joint venture attempted to raise capital by listing itself on the Tokyo Stock Exchange. A visit to the facilities of the joint venture by the head of the Exchange was a condition for listing. The Exchange official would not agree to the visit because he felt "it would be a loss of face if he were to attend a formal ceremony at a U.S.-managed plant" [Oppenheimer, p. 99]. In spite of the fact that there actually was only one U.S. manager on the Board of Directors, the joint venture had to remove him before listing.

When a technology is targeted, protectionism is not far behind. Intel, the world leader in microprocessors for personal computer applications, dominated the market in Japan until NEC introduced an 8080-type microprocessor into volume production. A coinventor of the monolithic IC, and Vice Chairman of Intel, testified during meetings on high-technology industries held by the Department of Commerce that "the market for 8080-type microprocessors collapsed in Japan while world demand continued" [Sanders, p. 21]. Despite the advantage of being the first entrant in the marketplace and of having economies of scale, NEC rapidly displaced Intel as the market leader.

The *Buy Japan* practice has been extended to optoelectronics technology for high-priority targeting and protecting because it not only encompasses a vital industry, but also a range of industries. Opto-electronics is an enabling technology for computers and communications. When Nippon Telephone and Telegraph (NTT) refused to purchase optical cable from Corning Glass in the early 1980s, it was to protect a burgeoning technology. At that time, Corning was a pioneering patent holder in optical fibers, and the first company to reduce fiber losses sufficiently for useful communications [Costa, p. 3], a 1970 milestone in the development of optoelectronics. Corning offered vastly superior, more reliable, and less costly optical fiber than was available anywhere. NTT purchased fiber only from domestic firms until the Japanese achieved rough technological parity and built up a large export capability. Only then did the Japanese make it appear that they were succumbing to U.S. pressure by giving contracts to U.S. firms.

The significance of this policy on optoelectronic technology is understood better when one realizes that optical fiber technology has been the driver for optoelectronic semiconductor technology in communications and computer networking. The properties of the optical fibers directly affect the development of the optoelectronic semiconductors. As noted earlier, optical fibers have so-called windows of transmission in which losses are minimized. Emitters and detectors are designed and manufactured so that their parameters are compatible with the transmission properties of optical fiber. Therefore, when Japan was sheltering optical cable technology, it also was insuring its options in optoelectronic devices, including OEICs. The Japanese already had targeted the OEIC in 1979 when it initiated a national optolectronic project.

This targeting of technology does not mean that the Japanese government does not target industries. When technology cannot be the instrument of protection, then the industry is protected openly, as is beermaking, which is sheltered by barriers preventing foreign firms from setting up production in Japan ["The billowing beer market," p. 32]. Japan protects the rice grower, not rice-growing technology, for political and cultural reasons [Reischauer, pp. 20–21, 302, 304]. The targeting of industries predated that of technologies. Japan copied U.S. products first before acquiring its own technological base. In the 1930s, Japanese suppliers of automobile replacement parts reverse-engineered Fords and Chevrolets under the protection of the Japanese government. Ford and General Motors were prevented from establishing manufacturing facilities in Japan during a critical growth period for the Japanese automobile industry [Wilkins, pp. 498–504]. The same situation prevailed in the electrical industry. U.S. multinational giants had invested in Japan in the 1920s, only to be squeezed out in the 1930s.

National Optoelectronic Projects

The implementation of the Japanese drive for optoelectronics preeminence began in 1979 with the establishment of the Optical Measurement and Control System project. Its research facility, the Optoelectronics Joint Research Laboratory (OJRL), began operations in 1981 [Tsang, pp. 5-1, 5-2].

The semiconductor members of the 16-firm OJRL are listed below:

Toshiba
NEC
Hitachi

Fujitsu
Matsushita
Mitsubishi
Oki

The development was carried out in the joint laboratory and in the individual member companies. Each company contributed research personnel to staff the joint laboratory. The researchers are permanent employees of the individual companies, and returned to their respective companies when the laboratory operations ceased in 1985. The laboratory, which consisted of about 50 technically trained people and less than ten administrators at its peak in 1984 and 1985 [Hayashi et al., p. 1434], was responsible for materials growth and processing. The problems of device manufacturing were left to the individual companies. The laboratory was funded directly by the government. The member companies funded their own manufacturing development. Each laboratory researcher transferred the laboratory results to his company. The total funding from the government was $77 million [Spicer, pp. 2-1–2-11].

The project, originally designed to advance technology in fiber optic local-area network (LAN) systems with the OEIC as the main development tool [Hayashi et al., p. 1451], enabled Japan to establish itself as the leader in optoelectronics technology and sales. Over 300 patents resulted from the project. An executive of one of the member companies stated that the OEIC was the most important device generated by the project [Howell et al., p. 121].

In 1986, the same year that the OJRL project was disbanded formally, the Optoelectronics Technology Research Corporation was founded. It is a ten-year project with a budget of $82 million and designed primarily to establish Japan as the undisputed leader in OEIC technology [Sakurai, p. 104]. The 13-member organization includes the following semiconductor firms:

Toshiba
Mitsubishi
Hitachi
Fujitsu

Matsushita
NEC
Sharp
Sanyo

The top six Japanese semiconductor firms are members of both projects. It is apparent that the results of the previous consortium proved satisfactory to these firms, and that they consider the OEIC to be a key device.

The project's research facility is the Optoelectronics Technology Research Laboratory (OTRL). As in the previous project, research in materials growth and processing is conducted at OTRL. Device manufacturing problems fall within the province of member companies. The success of the first optoelectronics joint venture, and the probable success of the second, are the result of a number of factors. The most important are the following:

1. The harmonious relationship existing between government and industry which has resulted in the identification and targeting of optoelectronics technology.
2. The efficient diffusion of technology because people as well as paperwork accompany the technology transfer from laboratory to member company.
3. The consensus and cooperation at the R&D level of firms that compete fiercely in the marketplace.

Encirclement

Encirclement is my term for a strategy to dominate enabling or support technologies critical to a specific technology or industry. Japan has used this strategy to achieve leadership in the semiconductor market. Japanese success in these technologies has made the U.S. dependent upon Japan for key semiconductor materials, piece parts, and manufacturing equipment. In 1975, the U.S. supplied nearly 80% of all semiconductor manufacturing equipment (SME) in Japan, with domestic firms supplying

the remainder. Ten years later, the numbers were nearly re-
versed [Berger, p. 26, July 22, 1985]. In 1982, the U.S. sold 62%
of the world's equipment for the front end of the manufacturing
line where the most critical processes take place, including
wafer steppers (lithography), ion implantation, epitaxy, and
sputtering. Japan's share of the market was 29%. By 1987, the
market virtually was split evenly between the two: 45% for the
U.S. and 44% for Japan [Rice, p. 28, May 15, 1989]. The bulk
of this market segment is in wafer steppers, used in the process
which determines the feature dimensions of an IC. In 1988,
Japan took over from the U.S. the world leadership in the total
SME market with a 49% share, compared to the U.S. share of
36% [Levine, p. 5]. Four of the top ten, including the first three,
were Japanese firms; six were from the U.S. [Mehler, p. 42]. In
1981, eight of the top ten SME suppliers were from the U.S.,
including the first seven. Japan held only the eighth and ninth
slots [Paul, p. 233].

SEMATECH, the U.S. memory consortium, has estimated
that in circa 1988, foreign firms sold from 70 to 95% of all of
the materials and equipment listed below purchased by U.S.
semiconductor companies, regardless in which country the
plant was located ["U.S. dependence on foreign suppliers," p.
1]. The compilation did not mention from which countries these
firms came, but the overwhelming majority are Japanese.[6] The
equipment and materials include the following:

Equipment

Stepping Aligners
Resist Processing

[6]This is based upon my analysis of the data supplied by research organizations to trade periodicals, or by comments from industry spokesmen and
observers. Europe and the ROW shared 15% of the world merchant SME
market in 1988. Most of this production was for domestic markets. Japanese
SME suppliers are considered to be aggressive marketeers with superior technology in equipment for high-volume semiconductor production lines [Lam-
mers, p. 1, Jan. 1, 1990]. It is not within the scope of this study to analyze the
semiconductor equipment and materials market. Two examples of the type
of data seen in trade periodicals follow: Nikon and Canon controlled 70% of
the stepper market in 1988 [VLSI Research Inc., p. 10]; Hoya Corporation of
Japan produces an estimated 55 to 65% of the mask blanks used in the U.S.
merchant market [Winkler, p. 26].

Scanning Electron Microscopes

Wafer Saws

Die Bonders

Tape Automated Bonders (TABs)

Molding and Sealing

Lead Trimming and Forming

Material

Silicon Wafers

Mask Blanks (U.S. market only)

Sputter Targets Lead Frames

TAB Tapes

Molding Compounds

Ceramic Packages

Hybrid Packages

Bonding Wire

In addition to silicon, Japan controls the GaAs and InP material market. Sumitomo leads in technology and market position for both materials.

In order to demonstrate how critical the above technologies are to semiconductor fabrication, both electronic and optoelectronic, I shall describe briefly the role of the mask blank in the manufacture of ICs. Glass mask blanks are purchased from the vendor by either the semiconductor supplier for in-house photomask fabrication, or a vendor of photomasks to semiconductor suppliers. The mask designer furnishes computer-generated IC designs in the form of magnetic tapes to the photomask shop. The shop creates the photomasks by translating these IC designs into a glass negative (or positive), using an electron-beam generator to form complex patterns on the mask blank, which has been coated with chromium, silicon, or iron oxide. Up to 30 individual photomasks, each defining a different set of circuit features, may be generated for each IC design, depending upon the IC type. The semiconductor manufacturer images each mask pattern separately onto the semiconductor wafer, whose diameter can vary typically from 2 to 4 in. (GaAs), and from 4 to 8 in. (silicon). The semiconductor wafer has been coated with a layer of photosensitive material (negative or positive resist). A projection camera optically images a reticle pattern on the moving wafer in a step and repeat process until the entire wafer is exposed. Image fields may be as small as 1 cm². It is critical to reliability and device yield that the masks are aligned accurately, and that mask defects are minimal. Typically, feature dimensions must be positioned accurately to within 0.125 μm over the entire surface of the mask. Line widths must be accurate to within 0.1 μm.

An estimated 90% of the mask blanks used by U.S. semiconductor companies are supplied by Japanese firms. In addition, Japanese merchant vendors supply an estimated minimum of 70% of the world's noncaptive photomask market. U.S. semiconductor production would grind to a halt if Japan ceased delivering mask blanks for whatever reason: technological, political, or economic.

Patents

As noted in Chapter 5, the U.S. led in external patent applications from 1969 to 1982, although Japan rapidly was closing the gap. Patent activity is a tool of protectionism for the Japanese government, and a sales tool for Japanese manufacturers. Texas Instruments filed its first patent application in Japan for the IC in 1960. Twenty-nine years later, it finally was granted the basic patent [Zipser *et al.*, p. C8]. Although this is an extreme example, it does take an average of six–eight years to process a patent in Japan, compared to 18 months in the U.S.[7] In contrast to the confidentiality of U.S. patent applications, the Japanese government makes public patent specifications after 18 months of the initial filing. Competitors can inspect the application and raise objections.

In the U.S. semiconductor industry, trade secrets often are a preferred method of protecting technology. Small companies,

---

[7]There is anecdotal evidence that the average time to award a patent to a Japanese company is one-half that required for a foreign company.

particularly, do not have the capital or time to fight patent infringements. Allied-Signal Inc. of New Jersey received a Japanese patent in 1989, 12 years after its first filing in 1977. During this long delay, seven unsuccessful challenges were made by the Japanese ["After a 12-year wait, Japanese patent awarded," p. 42]. Silicon Technology Corporation of New Jersey is still awaiting approval for patent applications filed in 1979 ["Access Japan," p. 40]. U.S. firms perceive that the patent process in Japan is deliberately designed to give Japanese companies time to study the patents for *patent flooding* [Yoder, p. B4]. This is a tool to overcome basic patents by making slight changes in the process or product and patenting the same. The company with the basic patents has two avenues of action: cross-license the Japanese competitor, or fight a costly legal battle.

A large company has the capital to fight patent infringement. Corning Glass filed a patent infringement suit in 1984 against Sumitomo Electric in the Southern District of New York, and obtained an injunction in 1987 prohibiting Sumitomo from selling optical fibers in violation of Corning's patents. In 1989, Sumitomo settled with Corning for $25 million and other considerations while the court was deciding what damages Sumitomo would pay ["Corning and Sumitomo settle patent dispute," p. 10]. Sumitomo now can continue to manufacture optical cable at its facility at the Research Triangle Park (RTP), NC. Sumitomo has become an accepted member of the U.S. optoelectronics manufacturing community. In 1989, it formed a joint venture with AT&T to manufacture optical fibers at RTP. AT&T owns 51% and Sumitomo owns 49% of LITESPEC INC. ["AT&T/Sumitomo joint venture company created," p. 10]. AT&T owned a 25% share in Sumitomo until 1924 when it disposed of all its foreign interests.

Technology Transfer

Cash-rich Japanese electronics firms have found an inexpensive technique for the transfer of technology. Japan is weak in basic research. Most R&D activities are conducted within industrial laboratories whose publications are monitored closely by corporate headquarters. In the U.S., basic research is funded primarily by the government, and is performed by the universities. The results of this research are available to all.

For a relatively modest amount of endowment money, Japan obtains access to premier centers of optoelectronics research such as the M.I.T. Lincoln Laboratory. It is here that the laser diode was invented concurrently with IBM and General Electric, and where pioneering optoelectronics research continues, most recently with the first monolithic integration of GaAs and silicon to create a GaAs/Si OEIC [Choi *et al.*, pp. 500–502]. Other optoelectronic targets include the University of Illinois, which, on October 6, 1989, dedicated a $13.5 million Microelectronics Laboratory, containing a Center for Compound Semiconductor Microelectronics. The Sony Corporation donated to the University $3 million to endow a chair in honor of John Bardeen, the coinventor of the transistor ["Sony grants $3M for U.S. semicon R&D," p. 37]. Another member of the Illinois faculty is Nick Holonyak, Jr., a coinventor of the LED.

Japanese firms comb the world's technical literature for new ideas. Many of their engineers fluently read English-language technical journals. Japan has access to all that is written in the U.S., whereas it is doubtful if U.S. engineers see as much as 20% of what is published in Japan.

Japanese semiconductor firms spent a considerable amount of time and money to send engineers to visit U.S. firms in the 1960s and 1970s. The stereotype of the Japanese engineer with a camera slung around his neck is not a false picture. In many cases, visits to U.S. firms were fruitful for the Japanese. Even if they could not take pictures of secret company processes and equipment, Japanese engineers learned a great deal from viewing the organization, layout, and routing on a factory floor.

Another source of technology for Japanese firms has been the U.S. federal laboratory system. The Federal Laboratory Consortium, established in 1974, was designed to assist universities and firms to utilize the resources of the 700 plus U.S. government laboratories and research centers around the country [Schneiderman, pp. 33–36]. More inquiries came from foreign sources than from U.S. sources. Japan, the U.S.S.R., and the

People's Republic of China have been leading customers of this service.[8]

Many U.S. manufacturers resent that the Japanese obtain royalty-free technology from the U.S. However, funding research, pursuing the U.S. technical literature, and obtaining information from U.S. government research laboratories are perfectly ethical and legal activities. They are better methods of transferring technology than the illegal methods sometimes used by Japan in the past. The best known example of this is the attempt by Fujitsu to illegally clone IBM personal computer software. Fujitsu paid IBM a total of $833 million over several years for this activity.[9]

This is not to say that Japan has not paid for technology. It is estimated that Japan spent a total of about $9 billion from 1950 through 1978 to accumulate technology from abroad [Abegglen and Etori, p. J20]. When you consider that U.S. industry spent about $74 billion for R&D in 1990 [Studt, pp. 40–44], the 30-year expenditures by Japan to close the technology gap was a bargain.

[8]Few large U.S. companies have used this source of information because there was no patent protection for products that they might develop from the research. Small U.S. companies could not cope with the government bureaucracy. The Federal Technology Transfer Act of 1986 alleviated some of the problems by allowing the laboratories to enter into cooperative research agreements, and to negotiate patent licensing agreements [Karlin, pp. 32–33]. NASA publishes a monthly digest on R&D results for which further information can be obtained either from the reporting entity or from the National Technical Information Service (NTIS), Springfield, VA. There is a considerable amount of data reported on optoelectronics R&D. For instance, researchers from NASA's Jet Propulsion Laboratory described a proposed random-access-memory (RAM) addressing system in which optical interconnects are used to relocate and isolate clock signals [Johnston *et al.*, pp. 32–33].

[9]In the 1970s, during a period of exponential growth by Japanese IBM-compatible manufacturers, IBM took no legal action against Japanese firms for software copyright infringements. In 1979, Fujitsu eclipsed IBM Japan as the leader in total computer sales in Japan. In 1982, IBM sued Fujitsu, Hitachi, and Mitsubishi Electric. In 1983, the three companies settled for $100 million. Fujitsu's share was one-third of the total. All agreed to a periodic IBM inspection of their operating-system software [Berger, p. 41, May 26, 1986]. Fujitsu continued to clone IBM software. The American Arbitration Association rejected Fujitsu's claim that its operating system was developed independently. Independent observers agree that the Fujitsu claim was not valid.

Japanese semiconductor firms smooth the path of technology transfer by their close attention to and cooperation with customers. Trade and professional associations are used as tools to obtain market feedback on present and future products.

Optoelectronic Industry and Technology Development Association (OITDA)

OITDA is one of the most powerful industrial technical organizations in the world, much more influential than any similar U.S. organization because of its emphasis on product development and applications [Forrest, pp. 73–74, Oct. 1989]. OITDA was founded in 1980 by eleven major Japanese corporations to promote research and development in optoelectronic, electrooptic, and optical technology. By the end of 1983, the association had 148 member firms. Nine of the founding firms are members of the MITI-sponsored Optoelectronics Joint Research Laboratory. Membership in OITDA is open to all companies, domestic or foreign, but only Japanese companies or Japanese subsidiaries of foreign companies are members. The activities of OITDA are monitored by committees and subcommittees and include the following.

Market and Technology Forecasting

- Research on innovative trends in domestic and foreign optoelectronic, electrooptic, and optical technology.
- Trends in systems employing such technologies.

Measurements and Standards

Data are collected, analyzed, and distributed on a broad spectrum of devices, equipment, and systems in all environments: land, sea, air, and space. Monthly reports and seminars are given to provide timely information. Detailed statistical analyses are published annually. All publications are in Japanese, except for an annual Activity Report that is published in English.

OITDA also sponsors projects that are assigned to individual members.

OITDA sponsors InterOpto, the premier optoelectronics components and systems exhibit in the world for products still in the R&D phase [Forrest, pp. 21–23, Oct. 1989]. In 1989, 90 000 people attended InterOpto, compared to the "record" crowd of 5000, including 1800 exhibitors, at the Optical Fiber Communications (OFC) conference, sponsored by the Optical Society of America (OSA) and the Institute of Electrical and Electronics Engineers (IEEE) ["Record crowd at OFC reflects surge in sales," p. 8]. The OFC bills itself as "the major North American conference on the technology and use of optical fibers for communications and related systems" [OFC '90, p. 7].

Although intended to be an international event, InterOpto is a showcase for Japanese technology. It is applications-oriented. Exhibitors use the show to obtain reaction from potential customers concerning the viability of their products. Suppliers will introduce products or modify product designs depending upon market feedback. The major Japanese optoelectronic suppliers redirect their development activities according to this feedback. A great deal of OITDA's influence originates from markets that the U.S. is losing (automotive) or has lost (consumer) to Japan. The loss of the latter has generated a ripple effect that has spread to the communications, computer, and defense industries. Displays, optical storage, and laser copy and facsimile are examples of markets now dominated by Japan because of its capture of the consumer market.

7

U.S. Optoelectronic Policies

Introduction

The U.S. has no national optoelectronic policy. Heilbroner and Singer succinctly characterized the attitude of the U.S. government towards business as being "chaotic and constrained, insecure and unfocused." American business is no better in its "narrow and antagonistic" view of government [pp. 349–350]. Government and industry do not identify, target, or protect a technology until it becomes an economic issue for a powerful special interest group capable of mobilizing opinion, or until it becomes a DOD protected item. Although individual interests are able to identify the OEIC as a critical technology, no national consensus exists. The DOD funds R&D programs in which optoelectronic device development is performed in aid of these programs. In order to examine the government's posture on optoelectronics, I shall review four reports that have been issued since 1985. Two are from the DOD, the third from the government agency concerned with measurements and standards, and the fourth from the Department of Commerce (DOC).

Planning Report 23, NBS, 1985

This is a previously referenced technology and assessment report on optoelectronics authored by G. Tassey, a Senior Economist with the then National Bureau of Standards (NBS), now the National Institute of Standards and Technology (NIST). Published in October 1985, this informative report states that the Japanese currently are leading in the worldwide race to capture the optoelectronics market. The U.S. can catch up if industry and government cooperate in the more rapid commercialization of optoelectronic technology. The Japanese Optoelectronic Joint Research Laboratory (OJRL) is cited as an efficient mechanism for achieving economics of scale and scope, and for transferring technology to industry in a timely manner.

The Tassey study estimated that over $1.0 billion will be spent annually by 1987 on worldwide optoelectronics R&D. Optoelectronics technology is deemed ready to *take off*. References are made to market projections by well-known technology and market research firms, ranging from $2.3 billion for optoelectronic devices in 1990 to $3.0 billion in 1990 for fiber optic components (optical fibers, transmitter/receivers, connectors). MITI is quoted as predicting that the aggregate world market for all optoelectronics, including devices, components, equipment, and systems, will range from $4.0 billion to $8.0 billion in 1992. (Since then, OITDA estimated that optoelectronics production for Japan alone was $15 billion in FY 1988 [Sakurai, p. 69], $17 billion in FY 1989, and that by the year 2000, Japan's total will increase to $110 billion, of which over $20 billion will be optoelectronic components ["Photonics in Japan: A booming business," pp. 50–51].)

The reason why the NBS sponsored this report was to obtain funding for measurement research. It projects an annual savings to the optoelectronics industry of $100 million–$200 million annually from its activities. This is based on past contributions to the productivity of the semiconductor industry.

The Tassey report does not identify the importance of OEICs for computer applications. It discusses the potential of OEICs in communications, but seems to favor lithium niobate (LiNbO$_3$) over III–V semiconductors because it will take ten years (from 1984) for the latter to reach the technological state in which the former is today. The author is astonished at the total commitment that the Japanese have to III–V OEICs [25], and Japan is "allocating most of its optoelectronic R&D to III–V semiconductors" [32]. The report describes the success that Japan has had in commercializing optoelectronic devices, but apparently discounts the fact that the Japanese already have targeted GaAs and InP over LiNbO$_3$.

It is beyond the scope of this book to analyze the relative merits of LiNbO$_3$ and GaAs. However, I question the reliance on LiNbO$_3$ because no viable laser has been made from this nonsemiconducting material. It is useful only as a substrate for hybrid OEICs. Hybrid devices inherently are less reliable and more costly than monolithic devices. The potential of LiNbO$_3$ lies in its use as an interim waveguide material for integrated optics in communications applications.

Why did this report value the potential of LiNbO$_3$ over GaAs and InP? The reason may be that one of its sources of technical information, the Batelle Memorial Institute of Columbus, OH, an independent, nonprofit, contract R&D organization which performs work for private and public sector clients, is more familiar with LiNbO$_3$ technology. Several years ago, it attempted to form a consortium to develop basic manufacturing technology for optoelectronic components. This consortium, entitled the Guided Wave Optoelectronic Manufacturing Technology Development Program, was to consist of 15–20 member firms, but only seven signed up: two connector firms, a semiconductor equipment manufacturer, and 3M, Litton Systems, ITT, and Hewlett Packard. Numerous reasons are given for its failure, but I believe that the primary one is the selection of LiNbO$_3$ as a production material. Optoelectronic semiconductor firms understand that the hybrid market is characterized by small production runs for high-performance applications. LiNbO$_3$ is not an appropriate technology for developing high-volume cost-effective manufacturing techniques. Batelle's choice may have been a reflection of its relative inexperience in the intricacies of real-time manufacturing and a misunder-standing of the future direction of the marketplace.

DOD Report on Electrooptics and Millimeter/Microwave Technology in Japan, 1985

A DOD technology team visited Japan from July 9–20, 1984 to establish a basis for the transfer of military technology from Japan to the U.S. [Department of Defense, May 1985]. The team found that Japan was lagging in defense system development, not unusual, but that it was very effective in transferring tech-nology from R&D to production. The team displayed a high interest in obtaining either design and test data, or production methods and know-how, or a supply source for ten technologies. All except one of the high-interest technologies are related to optoelectronics or electrooptics. Figure 7.1, derived from the report, lists the high-interest technologies.

Although the technology team visited all of the major OEIC developers, including NEC, Hitachi, Fujitsu, Mitsubishi, To-shiba, and Matsushita, it noted the monolithic integration of optoelectronic devices only at Matsushita. The term *OEIC* is not used anywhere in the report.

The stated goals of the DOD team were to: "investigate

Design/Test Data
  Optical Data Storage
  Laser Diode
  HgCdTe IR Detectors
Production Methods and Know-How
  Fiber Optic LANs
  GaAS Wafers
  Liquid Crystal Displays (LCDs)
  Electric Luminescent Displays
  III–V Materials
Potential Supply Source
  Imaging ICs
  Laser Diodes
  High-Density Memories

Fig. 7.1 Japanese Technologies of High Interest to the DOD in 1985.
Source: Department of Defense, p. iii, May 1985.

technological areas of interest; study mechanisms of technology transfer and modes of cooperation; and recommend structures for future reviews and further data exchanges" [p. ii]. If the DOD has obtained useful technology in the categories of design/ test data and production methods and know-how since the 1985 visit, these data have been hidden from merchant manufacturers. I question the effectiveness of any technology transfer in optoelectronics from Japan to the DOD by means of paperwork and working conferences only. The Japanese believe that a development cadre should be transferred with the technology, a concept well understood by U.S. policy analysts [McIntyre, p. 12], but not practiced enough by managers of large U.S. firms. Even this DOD report notes that the Japanese frequently stated that the transfer of talent is infinitely more important than the transfer of paperwork [pp. 1–7].

There also is an unstated reason for Japanese reluctance to transfer technology to the U.S. They do not desire to sell key technologies to their strongest competitor. Japan only wishes to supply products. All of the major OEIC developers, except Hitachi and Matsushita, have become semiconductor suppliers to the U.S. defense industry. In fact, NEC is the leading supplier of military millimeter/microwave semiconductors. Why should Japan transfer technology when the DOD team interpreted "modes of cooperation" to include the mission of researching out sources of supply? To transfer goods rather than technology is a perfectly rational stance for Japan to take.

The technology team visited Japan on the premise that the U.S. is not the only source for advanced technologies. This is a correct posture. The DOD should investigate key technologies developed by U.S. allies. However, it appears from my reading of the report that the team digressed from its mission when it considered Japan as a supply source for products that can be purchased domestically.

Figure 7.2 lists the products that the DOD team considered for potential procurement from Japan. All are optoelectronic devices or materials, except for high-density memories and microwave and millimeter-wave components. There are bona fide American suppliers for all of these products. Why is it necessary to purchase them from Japan? The necessity for asking this question is ironic when you consider that a former Assistant

Imaging ICs
Hybrid OEIC Photoreceivers
Laser Diodes
Avalanche Photodiodes (APDs)
Optoelectronic Materials
High-Density Memories
Microwave and Millimeter-Wave Components
GaAs Wafers

Fig. 7.2   DOD Procurement Wish List of Optoelectronic and
Microwave Products from Japan in 1985.
Source: Department of Defense, May 1985.

Secretary of Defense for Production and Logistics deplored the
U.S. military dependence on foreign parts when he was hired
in 1987. The mission of this former Executive Director of Purchasing and Material at General Motors was to plan a strategy
for "strengthening the U.S. manufacturing base in key industries" [Rothschild, p. 4]. He noted that there was no DOD control system to identify foreign-sourced parts.

The Defense Science Board Task Force on Semiconductor
Dependency reported in 1987 that reliance on semiconductors
from Japan threatens U.S. national security interests [Vogel, p.
26]. A government–industry survey, released in 1989, quoted
the top Air Force General for Logistics as saying that from 1982
through 1987, the number of suppliers to the military had decreased from 118 000 to 40 000 ["Survey warns defense base
eroding," p. 56]. The respondents to the survey were concerned
with the loss of subcontractors who furnish much of the critical
technology and innovation in U.S. weapon systems.

According to the Logistics Management Institute (LMI), a
government research center, the DOD procures 30% of its semiconductor requirements from foreign sources [Rayner, p. 84,
Jan. 23, 1989]. NEC and Fujitsu are key suppliers of GaAs microwave transistors, of which many have proprietary specifications. Mitsubishi and Toshiba are leading suppliers of memories to the DOD. Mitsubishi was the 31st largest electronics
supplier to the DOD in FY 1988, with sales of $692 million
[Callan, p. 63].

On the other hand, U.S. suppliers of key device technologies stand no chance of obtaining any significant business from the Japanese military. In FY 1984, the Japan Defense Agency spent $28 million on foreign electronic products out of a total budget of $11.7 billion [Blustain and Polishuk, p. 330]. None of the $28 million was spent on optoelectronics. By FY 1989, Japan's defense budget had climbed to $29 billion [Vogel, p. 22]. Japan purchases practically no semiconductors from foreign sources for weapons systems.

Procurement improves production yields, lowers unit costs, advances technology development, and reduces foreign device dependence. There are few military applications that afford the economy of scale which optoelectronic suppliers need for massive advances in cost-effective production. Experimental development without proving results through real-time production is inefficient. The failure of the DOD to unreservedly support the domestic semiconductor industry through purchases of standard commercial devices has been a factor in DOD problems in weapons reliability and deployment cycles, and in reducing the competitiveness of the U.S. semiconductor industry.

DOD Critical Technologies Plan

The National Defense Authorization Act for FY 1989 requires that the DOD annually identify to Congress those technologies that will ensure the long-term qualitative superiority of U.S. weapons systems. The idea is to integrate critical technologies into the DOD science and technology program. These critical technologies can be in any stage of research or development. New technologies can be added annually.

The DOD selected 22 critical technologies for its first submission in 1989. Twelve of them involve optoelectronic semiconductors. While no specific reference is made to OEICs in the entire report, descriptions of several applications imply the use of OEICs. These descriptions cite the need for the following:

Optical coupling and routing
Optical memory

Fiber optic LAN

Optical sensor

Figure 7.3 lists the critical technologies involving optoelectronic semiconductors, and the DOD's opinion as to who is
leading in each technology. Japan leads or shares the leadership
with the U.S. in the two enabling technologies that have the
most pervasive influence on the remaining technologies and on
most U.S. weapons systems. They are microelectronic circuits
and fabrication, and the preparation of GaAs and other compound semiconductors.

In microelectronic circuits, the U.S. leads in microprocessors, and Japan in memories. Japan is the world leader in supplying GaAs and other compound semiconductors. In micro-
circuit fabrication, Japan is ahead in most supporting
technologies, including basic materials, substrates, packaging,
and mask blanks. Because of its superiority in microprocessors
and software and its lead in integrating military systems, the
U.S. is ahead in the remainder of the technological areas, except

Microelectronic Circuits and their Fabrication	U.S./Japan
Preparation of GaAS and Other Compound Semiconductors*	Japan
Parallel Computer Architectures	U.S.
Machine Intelligence/Robotics	Japan**
Simulation and Modeling	U.S.
Integrated Optics	U.S./Japan
Fiber Optics	Japan
Sensitive Radars	U.S.
Passive Sensors	U.S.
Automatic Target Recognition	U.S.
Phased Arrays	U.S.
Data Fusion	U.S.

* GaP, GaSb, InAs, InSb.
** U.S. leads in computational capabilities.

Fig. 7.3 Critical Military Technologies Involving
Optoelectronics.

Source: Department of Defense, Critical Technologies Plan,
Mar. 15, 1989.

machine intelligence and robotics, which the Japanese have targeted for their industrial programs. Most of these military applications use technologies that can be converted to commercial use with appropriate modification and varying difficulties. For example, parallel computer architecture is being developed intensely by industry as a key to increasing computer throughput. Optical coupling and routing are technology tools in the drive to produce computer systems that are capable of trillions of operations per second (teraops). Automatic target recognition involves computer architecture, software, and high-speed signal processing. Data fusion requires advanced data processing techniques for a wide range of applications from battle management to integrated cockpit displays.

The second annual Critical Technologies Plan, published on March 15, 1990, selected 20 critical technologies. All of the critical technologies in Fig. 7.3 were contained in the second plan. The nomenclature and definitions were changed slightly because Congress asked that they be placed in context with the DOD's total Science and Technology Program. The technologies were divided into three priority groups. Top priority was given to the most pervasive technologies. Grouped in alphabetical order, they include:

Composite Materials
Computational Fluid Dynamics
Data Fusion
Passive Sensors
Photonics
Semiconductor Materials and Microelectronic Circuits
Signal Processing
Software Producibility

The plan describes photonics as "the product and process technology for devices that use light (photons) and electronics (electrons) to perform functions now typically performed by electronic devices" [p. A-65]. The OEIC specifically is cited as one of the critical technology challenges in photonics. The report states that "the use of OEICs would also extend weapons capabilities in the areas of automatic target recognition, state-of-health monitoring, and detection avoidance." Also noted is the potential benefit of photonics R&D to the industrial base, particularly in the area of high-speed computing.

Critics of the military's R&D programs contend that the Japanese are leading in technologies critical to defense because the DOD does not have a coordinated strategy for technology. A study of the Critical Technologies Plan by the Office of Technology Assessment (OTA) states that planning starts as a bottom-up process in each service, but never develops into a rational strategy [Agres, p. 20]. The OTA report cites the long time required to diffuse technology into military systems, and the inability of the DOD to absorb commercial technologies.

In the U.S. military, the integration and rationalization of military R&D is a political game among the services. Pork barrel politics flourish. States with the greatest political clout can affect military R&D decisions. There is little attempt to cooperate with industry in targeting and performing R&D on dual-use technologies that are critical to the defense industry, but whose largest markets are commercial. The Office of Technology Assessment reports that DOD dependence "is most pronounced in . . . semiconductors, computers, machine tools, structural materials, and optics" [U.S. Congress, p. 15].

Defense Advanced Research Projects Agency (DARPA)

DARPA is the primary research and development arm of the DOD, and the lead agency for semiconductor R&D. It was established in 1958 to sponsor applied research and experimental development in technologies that are too risky for conventional funding, but offer exponential leaps in military weapons performance if successful. Until 1988, the DOD spread responsibility for its semiconductor programs among the Office of the Secretary of Defense (OSD) and the three branches of the armed services. The Army, Navy, and Air Force supported semiconductor programs directed towards their respective military missions. DARPA fell under the responsibility of the OSD.

In 1988, the scope of DARPA was broadened to include all

DOD semiconductor programs, with the exception of certain "black box" projects funded by the Army Security Agency (ASA) and the Central Intelligence Agency (CIA). DARPA is comprised of ten separate technology offices. The Microelectronics Technology Office is responsible for "component and manufacturing technologies necessary to information processing" [Galatowitsch, pp. 23–41]. Among the programs are those involving optoelectronics and optical processing, and digital gallium arsenide (GaAs) microelectronics. A goal of the optoelectronics program is to persuade industry to accept and use optoelectronic devices.

DARPA's major semiconductor responsibilities now include the Very High Speed Integrated Circuit (VHSIC) and Microwave/Millimeter-Wave Integrated Circuit (MIMIC) programs, and the government's participation in the Semiconductor Manufacturing Technology (SEMATECH) program, a cost-sharing consortium of semiconductor manufacturers whose first goal is to regain U.S. leadership in semiconductor manufacturing technology [Department of Defense, May 12, 1988]. The DOD contributes one half of the annual funding of $200 million for SEMATECH, and all of the funding for the MIMIC and VHSIC programs. The additional responsibility for these programs and others means that DARPA's role has been expanded to include the development of specific products for specific weapons systems.

The VHSIC program was designed to obtain advanced silicon ICs for insertion into existing weapons systems and for the design of new systems. This ten-year program, which started in FY 1980 and was to be completed by the end of FY 1989 (September 30, 1990), was designed to cut from 12 years to two years the time that it takes to introduce a microelectronic technology into a weapons system [Department of Defense, p. ii, Dec. 31, 1987]. No optoelectronic device came out of this billion dollar program, although some effort was extended to investigate the use of optical fibers to replace off-chip electrical connections. Useful technical data from this program are not available to anyone except the defense community on a "need-to-know" basis [84]. In the journals, Motorola did publish two papers concerning its work on silicon optoelectronics. One described a monolithic silicon OEIC receiver in which a p-i-n photodiode is monolithically integrated with an amplifier/limiter circuit for operation in short-haul applications at data rates up to 500 Mb/s [Hartman *et al.*, pp. 729–738, Aug. 1985]. The other described a coupling and packaging technique for a silicon OEIC in which the input is implemented with optical fibers, and the output with standard wire-bond interconnects [Hartman *et al.*, pp. 73–82, Jan. 1986]. The authors note that III–V semi-conductors offer higher performance levels than silicon.

The MIMIC program is designed to accelerate the development of GaAs microwave/millimeter-wave ICs (MMICs) and the insertion of these analog ICs into radar, electronic warfare, communications, and smart munitions systems [Department of Defense, May 20, 1988]. The DOD awarded initial contracts in FY 1987 for this seven-year program. Total estimated funding is $600 million. No optoelectronic devices have been specified in MIMIC.

SEMATECH is a semiconductor memory consortium of 14 large billion-dollar semiconductor and computer corporations [Harvard Business School, p. 24], with the exception of two semiconductor firms whose sales are in the $400 million range. The computer firms have captive and/or merchant semiconductor operations. The fee structure for membership makes it proportionally more expensive for a small company to join. This is unfortunate because it is the small companies that are the device innovators. They do not have the capital to develop manufacturing capabilities that will give them the staying power in the marketplace.

SEMATECH is a consortium with problems. It took a year to appoint a Chief Executive Officer [Van Nostrand, p. 1]. The operating plan was to develop high-density memories. Because its timetable has been slipping, because the Japanese have introduced these memories more quickly than SEMATECH anticipated, and because the U.S. rapidly is losing ground in semiconductor manufacturing equipment markets and technologies, the emphasis has been shifted from device to equipment technology. There is no provision in this consortium for OEIC memories, although this is a promising way to leapfrog Japan in memory technology. The Office of Technology Assessment notes that SEMATECH is criticized as a short-term attempt to solve a specific device problem. It is not prepared to "maintain the state-of-the-art capacity if global market forces push subsequent generations of equipment into gallium arsenide or optically based technologies" [U.S. Congress, p. 40]. SEMA-TECH is not concerned with long-range technological planning, only with stopping the present hemorrhaging in the market-place.

Analysts argue that government R&D and procurement helps U.S. companies in the commercial marketplace [Flamm, p. 80]. This was more true in the 1950s and early 1960s when military and space applications accounted for a large share of semiconductor sales, ranging from 38 to 45% [Roessner, p. 234]. Government funding did contribute mightily to semiconductor advances during this period. The government not only was a large market, but was willing to take standard commercial product, in fact, anything that the firms could supply. But market conditions have changed. Estimates of semiconductor sales to government contractors and agencies now range from 5 to 8%, depending upon the source of data. Defense goals in semiconductor technology now are different from those in the commercial market. Military designers prefer to write their own semiconductor specifications rather than specifying standard commercial product. DARPA funds a considerable amount of R&D for semiconductor technology, which may be useful for the military's mission, but which is not cost-effective for the commercial market.

As noted in Chapter 4, the U.S. government spent $454 million in 1987, not including tax credits to industry, for semiconductor R&D. Of this total, 6.5% was expended on optoelectronic devices, of which 51% went to photovoltaic research. Photovoltaic cells have a long-range potential, but they represent a minor fraction of the current optoelectronic semiconductor market, and will not be a major market factor for the rest of this century. HgCdTe infrared (IR) detectors accounted for 8% of total R&D expenditures for optoelectronic semiconductors, and at least 19% of the DOD portion. HgCdTe IR detectors operate in a cooled environment and have a spectral range (8000–12 000 nm) beyond the requirements of the commercial market, except in highly selective applications. These detectors have not been integrated monolithically with electrical functions on a single substrate, as have GaAs devices. Although the DOD is funding development to integrate HgCdTe IR detectors with silicon circuits, the R&D costs associated with these efforts are not warranted commercially based upon the predicted size of the market and projected device costs. HgCdTe IR detectors are high-performance devices for applications where cost is of no consideration. The IR detectors apparently performed well in the Gulf War in a variety of thermal imaging applications. The television medium graphically presented their precise targeting capability in airborne systems. An array of cooled HgCdTe detectors in the Bradley tank [Kales, p. 121, Nov. 1990] seems to have contributed to its superiority over Soviet-made Iraqi fighting vehicles.

In 1988, Bell Laboratories reported the development of GaAs IR detectors operating in the HgCdTe spectral range (8000–10 000). GaAs is a more stable material and more cost-effective to manufacture. Military designers use HgCdTe detectors because they are extremely sensitive. However, overall system sensitivity is improved materially with monolithic integration. Reliability is increased, and weight and size are reduced.

The funding of HgCdTe IR detectors is another example of the divergence of defense and commercial goals in the development and application of a device. The DOD is seeking performance ahead of all other considerations. Commercial manufacturers must balance performance with other considerations, including cost and manufacturability.

The U.S. government spends a considerable amount of its annual semiconductor R&D budget on radiation-hardened (rad-hard) device technology. This is a legitimate concern for the DOD and DOE. Hazardous radiation is found in military and space environments. Cosmic radiation effects on semiconductors were first noted in the early 1960s. Catastrophic failures and degradation effects can be caused by the explosion of a nuclear device. The materials in high-reliability ceramic IC packages used by the military can contain radioactive contaminants. Radiation problems have increased as the density levels of ICs have increased. Feature dimensions of ICs have become so small that it is possible for a circuit to have its state altered by a single ionizing particle.

The U.S. government spent $98.6 million in FY 1987 on rad-hard semiconductor R&D. This represents 21.7% of the semiconductor R&D budget compared to 6.5% for optoelectronic semiconductors. The market for rad-hard semiconductors worldwide was less than 1% of the total semiconductors sold in 1987 [compiled from Rayner, p. 22, Sept. 18, 1989, and Instat/SIA, p. 29]. Because the DOD bars U.S. suppliers from exporting semiconductors hardened above a certain radiation level [Rice, p. 18, July 1, 1989], the total available market for U.S. suppliers is reduced to less than 0.6% of the worldwide market.

While important to the DOD and DOE, radiation hardening makes little contribution to U.S. semiconductor competitiveness. Radiation hardening requires process steps which decrease yields and degrade performance levels, while increasing unit costs. Radiation problems resulting from manufacturing can be corrected by process changes. Commercial producers do not use much ceramic packaging. Except for military, space, medical, and rugged industrial applications, there are few significant uses for rad-hard devices. Most semiconductors operate in relatively benign environments.

Federal R&D expenditures on semiconductor programs have contributed very little to U.S. competitiveness in optoelectronic semiconductors. The VHSIC program was for silicon devices only. The MIMIC program does not include OEICs. SEMATECH is a floundering program for a lost silicon memory market. U.S. military semiconductor R&D programs offer no help for GaAs OEIC suppliers. This is unfortunate because GaAs OEICs will exhibit a greater tolerance to total radiation dose effects than will silicon OEICs. There are considerable data showing that GaAs ICs are superior to silicon ICs in this property [Sleger et al., pp. 85–86]. GaAs OEICs could solve some of the problems associated with ionizing radiation.

Department of Commerce Report on Emerging Technologies

The Department of Commerce (DOC) is not concerned with radiation resistance or other strictly military requirements. In 1990, three years after its first report, the DOC issued its second report on emerging technologies in "promising fields with large potential economic impact" [U.S. Department of Commerce, Spring 1990]. The report identified 12 emerging technologies with a "combined U.S. potential of $356 billion in annual product sales by the year 2000 . . . and a world market approaching $1 trillion" [p. V]. These technologies are grouped into four categories: materials, electronics and information systems, manufacturing systems, and life sciences applications. The category of electronics and information systems comprised about 45% of the total annual U.S. sales by the year 2000, the largest of the four categories.

The electronics and informations systems category includes optoelectronics, advanced semiconductor devices, digital imaging technology, high-density storage, and high-performance computing. Optoelectronics is forecast to annually generate $4.6 billion in U.S. sales and $10.8 billion in world sales by the year 2000. However, a review of all of the technologies indicates that the potential market for optoelectronics is understated because many of the other emerging technologies will require optoelectronic devices, including OEICs, for their successful application. Figure 7.4 lists the 12 technologies in the electronics and information category, and indicates where the OEIC can make a valuable contribution.

A closer examination of the forecast figures for optoelectronics shows that they are derived from a 1988 report of the DOC which covers the fiber optic communications market only [U.S. Department of Commerce, International Trade Administration, p. 25]. The applications are all in communications, except for those in military guidance and sensing, and in medical systems. No mention is made of the potential of optoelectronics for high-performance computer applications. This is understandable because the forecast estimates are based on responses received from questionnaires sent in 1987 to "nine research firms or major U.S. companies in fiber optics" [p. 25]. In 1987, most major U.S. corporations in fiber optics were thinking that the best markets were in communications, not high-performance computing.

As noted earlier, Japan's Optoelectronic Industry and Technology Association (OITDA) forecast that in the year 2000, domestic optoelectronic production will be over $100 billion.

	OEIC
Advanced Materials*	
Advanced Semiconductor Devices	X
Artificial Intelligence	X
Biotechnology	
Digital Imaging Technology	X
Flexible Computer-Integrated Manufacturing	
High-Density Data Storage	X
High-Performance Computing	X
Medical Devices and Diagnostics	X
Optoelectronics	X
Sensor Technology	X
Superconductors	

* Ceramic, composites, alloys, polymers, diamond thin films, surface-modified materials, and biomaterials.

Fig. 7.4 OEICs in Emerging Technologies.
Source: Author's estimate based on: U.S. Department of Commerce, Technology Administration, *Emerging Technologies*, Spring 1990.

OITDA anticipated that the OEIC will be in wide use by the year 2000.

The DOC report compares the relative competitive standing of the U.S. in the 12 technologies with Japan and the European Community (EC). In optoelectronics R&D, the U.S. is even with Japan and is holding its ground. But in the introduction of optoelectronics products, the U.S. is behind Japan and is losing ground. Compared to the EC, the U.S. is ahead in product introduction, and is holding its ground. In optoelectronics R&D, the U.S. is even with the EC and holding its position.

In none of the technologies in the DOC report is the U.S. gaining ground. In product introduction the trend is against the U.S. in nine of the 12 emerging technologies. The trend line for the U.S. in R&D is downward in five technologies. Furthermore, in advanced semiconductor devices, the U.S. is behind Japan in product introduction and is losing ground. In advanced semiconductor R&D, the U.S. is even with Japan and holding its own.

The report's outlook concludes that [p. 25]

> the economic growth of many nations, especially that of the United States, has been based on the development and successful introduction of emerging technologies . . . Lately U.S. industry has been unsuccessful in capturing the majority of benefits from emerging technologies . . . trading partners have demonstrated substantial economic growth through the marketing of products based on U.S.-developed technologies.

The report cites the barriers that must be overcome to reverse the U.S. technological decline [p. 15]. However, these barriers are general in nature and have been thoroughly aired in the past. No consensus for action has been reached to date on these obstacles, euphemistically referred to as opportunities for change in the report, because they are viewed differently by many powerful adversarial political and economic interests.

Proposed U.S. Industrial Optoelectronic Consortium

Eleven years after the first Japanese national optoelectronics project began, concerned U.S. firms attending the Optical Fiber Communications (OFC) Conference in January 1990 started serious deliberations for a consortium of U.S.-based optoelectronics companies. Subsequently, at a meeting sponsored by the Department of Commerce, about 30 manufacturers explored ideas presented by the David Sarnoff Research Center, Princeton, NJ, a subsidiary of SRI International, Menlo Park, CA. The Sarnoff Research Center, formerly the RCA Research Laboratories, was given to SRI in 1987 by General Electric, which had acquired the Laboratories in 1986 when it purchased RCA. About 10% of the Laboratories' activities were focused on semiconductor R&D.

The Sarnoff Research Center has proposed a six-year plan, costing $155 million, to develop and prove the manufacture of OEICs based upon silicon VLSI (very large scale integration) chips with input and output interconnects implemented with laser diode and photodiodes [Robinson, p. 6, Mar. 6, 1989]. The center has developed a laser diode technology, GSE (grating surface emitting laser), which can be used in fabricating OEICs. In July 1989, six firms with semiconductor operations, including Analog Devices, AT&T, Harris, Hughes, Motorola, and Tektronix, met at the Sarnoff Research Center for further discussions. Motorola was the only one of the top five U.S. semiconductor manufacturers to attend. To date, the consortium has not been established.

The Sarnoff Research Center has an excellent reputation for innovative research, and is a pioneer in the development of complementary metal–oxide–semiconductor (CMOS), the leading silicon IC technology. It is a logical choice to coordinate the activities of a semiconductor consortium. However, the integration of GaAs on silicon (GaAs/Si) means that the consortium will be developing concurrently for manufacturing two technologies: the OEIC and GaAs/Si. Researchers at the Center understood in 1987 the benefits and problems involved in GaAs/Si development [Menna, p. 59]. The Center realizes that "electronics is fundamentally superior to optics for computing, but for communications and/or sensing, optics is superior," but predicts that the GaAs/Si OEIC will not emerge from the laboratory until the end of the century [Ettenberg, pp. 149–150]. The U.S. cannot wait that long to introduce the OEIC to the marketplace. It should enter the market with a GaAs OEIC.

It is beyond the scope of this work to compare the advantages and disadvantages of silicon and GaAs as OEIC substrates. However, one might question the wisdom of trying to do too much pioneer development when the goal is to create a manufacturable and marketable product. Japanese semiconductor firms have proven that small steps rather than large leaps are a faster path to the marketplace. One reason that the U.S. lost the memory market is because U.S. memory leaders spent too much design time trying to add as many features as possible to the 64-kb DRAM. The Japanese jumped into the market ahead of the Americans by designing a simple reliable device. The market was eagerly awaiting this higher density memory. Customer circuit designs were established by the time U.S. suppliers introduced their memories, replete with more "bells and whistles" than Japanese devices. The customers wanted rapid deliveries of large volumes of memories. The Japanese understood this, and reacted accordingly. They quickly dropped unit prices so that it was cost-effective for the market to switch from the 16-kb memory to the 64-kb device. American suppliers were entranced with the technology. The Japanese moved according to the market demand. They emphasize the commercial development of technology rather than the discovery of specific devices [Starling, p. 195].

R&D is being conducted by U.S. and Japanese companies on GaAs/Si ICs as an alternative to GaAs ICs. Texas Instruments has developed a GaAs/Si 1-kb memory [Shichijo *et al.*, pp. 121–123]. M.I.T. Lincoln Laboratory, with funding from DARPA, has pioneered in GaAs/Si R&D and has grown laser diodes on GaAs/Si substrates [Windhorn *et al.*, pp. 157–164, 1986]. Kopin Corporation [1987] currently is the only reliable U.S. supplier of GaAs/Si wafers. Kopin was founded in 1984 by the former head of the Electronic Materials Group at Lincoln Laboratory. However, GaAs/Si OEIC device development lags behind that of the GaAs OEIC. American engineers understand the difficulties involved in integrating dissimilar materials. GaAs already is an established material with a mature device process. The U.S. is trailing Japan in OEIC product development. Any U.S. industry consortium should keep it simple and concentrate first on bringing to market GaAs OEIC technology.

U.S. industry lags behind Japan in establishing precompetitive consortia to develop critical technologies. On the other hand, the U.S. university research system is the best in the world. Backed by the defense establishment, the U.S. continues to outpace Japan in the quality of basic optoelectronics device research. DARPA has awarded a contract for about $12.5 million to a consortium of five universities, led by the University of Southern California (USC), for the establishment of the National Center for Integrated Photonics Technology. Congress added these funds to the FY 1989 budget as part of a university R&D program in optoelectronics and photonics [Robinson, p. 1, Feb. 20, 1989]. Congress apparently has responded to the suggestion by the National Research Council for a national photonics project that includes industry, government, and academia [National Research Council, pp. 70–72]. The Director of the Center said that: "the Center hopes to advance significantly the state of the art in integrating optical and electronic devices on a single chip" [Santo, p. 35].

In addition, **DARPA** is establishing two other university centers, the Optoelectronics Center, a three-member consortium headed by Cornell University, and the Optoelectronic Materials Center, a five-member consortium led by the University of New Mexico. These two centers received approximately $6 million each. The funding for the centers may be spread over two–three years.

The **DARPA** Program Manager for optoelectronics and optical interconnects was interviewed by the *IEEE Circuits and Devices Magazine* [Fouquet, pp. 51–53]. The Manager stated that:

> The Military Critical Technologies List reflects the importance that DOD attaches to different technologies. I am especially pushing optical interconnects, optoelectronic circuits in III–V compounds . . .

The **DARPA** Manager is concerned about Japanese investment and advancement in the area of III–V compound semiconductor optic-related research. The purpose of the optoelectronic centers is "to get the universities to work closely with industry in order to get a good quick transition of their results into American industry." About two-thirds of **DARPA** funding is for research; the remainder goes for development.

During the past decade, federal funding for technology-related research has decreased in real terms. The Council on Competitiveness, a nonprofit, nonpartisan organization whose national affiliates include trade associations, professional societies, and research organizations with roots in business, higher education, and organized labor, has concluded that government R&D funding since 1980 reflects the low priority given to industrial development [Council on Competitiveness, pp. 13–14]. The Council reported in 1991 that: "in 1988, only 0.2 percent of the U.S. federal R&D budget was devoted to industrial development, compared to 4.8 percent for Japan and 14.5 percent for West Germany." Defense accounted for 65.6% of the U.S. federal government R&D budget, compared to 4.8% for Japan and 12.5% for West Germany. The report concluded that the U.S. is losing badly in gallium arsenide materials, integrated circuit fabrication and test equipment, and optical information storage, all optoelectronics-related technologies.

Optoelectronics Industry Development Association (OIDA)

In 1991, 11 years after the Japanese founded the previously described OITDA, nine U.S. optoelectronic suppliers and users formed OIDA, the North American counterpart to OITDA. Like OITDA, OIDA emphasizes the identification of technologies, products, and applications that will drive the optoelectronics market over the next 20 years. OIDA anticipates reaching its goals through a variety of means, including the "facilitation of pre-competitive R&D among members and government supported R&D," and the "coordination of relevant R&D programs at universities and national laboratories" [OIDA Brochure].

The charter members of OIDA include five firms with semiconductor operations: AT&T, GM/Hughes, Hewlett Packard, IBM, and Motorola. Hewlett Packard is the world's largest non-Japanese optoelectronics semiconductor supplier, estimated to rank fifth or sixth in sales in 1991, neck-to-neck with Sanyo, and significantly behind the four other leading Japanese manufacturers.

The other four charter members are AMP, the leading connector manufacturer, Bellcore, the research arm of the regional Bell telephone companies, Dupont, the major materials supplier, and GTE Labs, a leader in optoelectronic device research.

The influence of OIDA will depend significantly on how rapidly it can expand its membership. OITDA, founded in 1980 by 11 Japanese firms, increased its membership to 148 users and suppliers in three years.

8

What OEIC Policies Should the U.S. Adopt?

In Chapter 6, I outlined those policies that gave Japan its advantage in OEIC development. U.S. deficiencies in OEIC development were reviewed in Chapter 7. In this chapter, I shall recommend four policies that the U.S. should adopt in order to compete in the OEIC market.

Because the U.S. is over ten years behind Japan in organized planning for the commercial OEIC market, I have chosen policies which can be implemented easily and quickly. They are steps that the government, in cooperation with industry, can take within the existing U.S. social, economic, and political structure. There is no time to develop a grand strategy, which could take years to implement. The communications OEIC is emerging from the laboratory. The computer OEIC is only three years behind. The OEIC will be in common use by the end of the century.

Figure 8.1 outlines the following policy recommendations: establish a national OEIC foundry, procure OEICs for govern-

	Government Action	Government/ Industry Action	Policy of Japan
National OEIC Foundry		X	
OEIC Procurement	X		X
Export Decontrol	X	X	X
Technology Data Base		X	X

Fig. 8.1   Recommended OEIC Policies.
Source: Author.

ment applications from domestically based sources only, de-control semiconductor exports, and establish a technology data-base. The last three policies are followed by Japan. The first, the establishment of a national OEIC foundry, is designed to attack the greatest weakness in U.S. semiconductor competi-tiveness: its failure to convert technology into a marketable product in a timely manner. This is the only policy decision that will require a significant amount of funding. However, it is an investment with a cash return, aside from its strategic consid-erations. OEIC procurement and export decontrol require no appropriated funds. The establishment of a technology database will be a modest expense for industry and government.

Establish a Government-Subsidized OEIC Foundry

A foundry in semiconductor parlance is a facility to manufac-ture ICs that have been designed by a customer, usually another semiconductor firm with limited or no production facilities. The design and tooling are owned by the customer. A foundry can offer complete manufacturing facilities, or specific services such as wafer probing or packaging and testing. Foundries usu-ally are owned by large semiconductor firms that specify the manufacturing technology that they are offering. Capabilities advertised include the types of semiconductor material, usually

GaAs and/or silicon, and wafer sizes that can be processed, and the types of equipment available. Many small U.S. companies, rich in design talent but capital-poor, use U.S. and Pacific Rim foundries.

A total of 170 U.S. semiconductor firms were founded between 1979 and 1989 [Dataquest, p. 35]. About 60% of the existing start-ups do not own a wafer fabrication facility. Semiconductor firms that cannot forecast annual revenues above at least $100 million can ill afford the expensive manufacturing and test equipment required for a production facility.

Foundry operations are a profitable way for a richer company to utilize excess capacity and increase benefits accruing from economies of scale. A bonus is that the foundry acquires some new design and tooling know-how. Companies offering foundry services use the business "as a tap on leading-edge technology and markets" [Sack, p. 35]. In many instances, U.S. firms initiate the alliances, not because it is a strategically desirable move, but because they have no other choice if their product is to come to market. The Japanese and other Pacific Rim countries are most aggressive in foundry operations.

Many large U.S. semiconductor firms ignore the foundry market, thus exporting technology and jobs to erstwhile competitors and contributing to the decline of U.S. semiconductor competitiveness and to the increase in the trade deficit. It makes no sense for small U.S. firms to go offshore for production capacity. It helps the competition to lower unit costs and acquire new manufacturing insights. Many foundry agreements explicitly involve the transfer of technology. The major Japanese semiconductor firms are adept at trading production capability for technology.

The Computer Systems Policy Project (CSPP), an affiliation of 11 top U.S. computer manufacturers, concludes that the U.S. lags behind the world in manufacturing technology because of the high cost of construction for semiconductor production facilities. The Chief Executive Officers and the Chief Technical Officers of these companies noted that optoelectronics is one of the 16 technologies critical for maintaining U.S. leadership in computers ["CEOs: U.S. lags in fabs," p. 16].

The fastest and cheapest way to establish an OEIC foundry

is to make an existing government agency responsible for its planning and implementation. DARPA is the best choice because it has been involved with semiconductor programs for a long period of time, and it recognizes the need to support high-technology industries.

Congress recently expanded DARPA's role to include "collaborative research agreements and joint ventures with small R&D firms" [Congressional Budget Office, p, 107, July 1990], subject to certain limitations, including the provisos that the private sector partner will return DARPA's investment if the joint venture makes money, and that DARPA's investment will not exceed 50% of the total funding for the joint venture. The vehicle for this financing is a revolving fund authorized by Congress in the FY 1991 budget to which DARPA can contribute $25 million annually over a two-year period. This is double the amount authorized by Congress in FY 1990, despite the objections by the present administration. The funding for FY 1991 provides for $20 million of the $50 million total to be invested in optoelectronics and all-optical networks [Robinson, p. 16, May 6, 1991].

When DARPA, in 1990, exercised this authority allowed by Congress, its Director, who had been with the agency for 16 years, the last year as the Director, was transferred out precipitately to another DOD post [Robertson, p. 8, Apr. 23, 1990]. He had authorized an investment of $4 million in a small semiconductor firm founded in 1986 to supply gallium arsenide (GaAs) integrated circuits (ICs). Apparently, the administration viewed this encouragement of advanced semiconductor technology as an effort to set an industrial policy, something that is opposed vehemently by the administration. DARPA's goal in this investment was to assure that the industrial base needed by the DOD was strengthened by supporting an industry whose technology is critical to the success of DOD missions. Although the U.S. developed the first GaAs devices, it is falling behind Japan in both development and product introduction. What can be done to reverse this decline?

Reorganize DARPA into two sectors: *military* and *civilian*. The civilian sector will fund the foundry as a separate public sector corporation. This corporation will offer foundry services at a subsidized price to small U.S. companies, and at a market price to large U.S.-based companies.

Encourage U.S. start-ups to use the foundry; they are the most responsible for the introduction to the merchant market of new products. This is particularly true of the optoelectronics market. These start-ups are groundbreakers whose success depends upon market timing as much as upon technology. They do not have the staying power of the large, capital-rich, vertically and horizontally integrated Japanese firms whose profitability is not dependent upon devices, but upon equipment and systems. Japanese semiconductor firms are able to obtain cheap capital for device development and plant investment, not only because they are industry leaders, but because they are in a targeted high-technology field. The banks understand that the Japanese government is pledged to offer maximum assistance to semiconductor firms [Tooker, pp. 21–22].

A foundry can be established more quickly than a consortium like SEMATECH. Consortia are too expensive and time-consuming for small semiconductor firms. They cannot afford to contribute valued and scarce technical personnel, management time, or money to consortia that are dominated by large firms with vested interests and political clout in Washington.

An OEIC foundry will accrue a body of knowledge about manufacturing technology and organization, an area where U.S. semiconductor firms are weakest. This knowledge can be passed on to U.S.-based firms. A foundry will allow start-ups to concentrate on their major survival problem, which is how to sell the product. Most start-ups fail because of the lack of marketing expertise, not technological capabilities [Starling, p. 181]. Traditionally, U.S. high-technology factory start-ups in electronics tend to spend their capital on engineering, manufacturing, and marketing, in that order. Investment funds often are depleted by the time the marketing budget is implemented. The result is a start-up with top-grade engineers, but mediocre marketing people. Those entrepreneurs who understand the importance of marketing, but do not have the funds to invest adequately in all three areas are forced to turn over the manufacture of their product line to an outside firm.

A New Government Semiconductor Procurement Policy

*Require government agencies and defense
contractors to procure OEICs only from companies
with production and R&D facilities in the U.S.*

The Pentagon has undercut the domestic base in optoelectronics by importing optoelectronic semiconductors from such Japanese companies as NEC, Fujitsu, and Mitsubishi. The chances of U.S. firms regaining superiority in the optoelectronic marketplace are enhanced if the government procures OEICs from companies with U.S.-based production and R&D facilities. This policy gives U.S.-based companies a larger assured market base. Foreign companies with U.S. facilities may supply the government market as long as their U.S. facilities are not dependent upon foreign sources in any way.

A government policy of purchasing OEICs only from U.S.-based sources will help the small entrepreneur, the leader in the U.S. in the introduction of innovative products. There are all sorts of schemes afoot to help the small company. The latest is the proposed American Preeminence Act, an authorization bill passed by the House of Representatives. Among its provisions is one that authorizes a low-interest loan program for companies wishing to introduce their innovations to the marketplace [Barrett, p. 9]. This $10-million program, already threatened with a veto by the administration, is designed for small and medium-sized firms. This is a nice gesture by the House, but it will not help small firms in the semiconductor industry. If a company is successful in the market with its new products, more credit is needed. If a company is unsuccessful because its timing is poor or marketing strategy flawed, more credit is needed. A committed source of funds is necessary to see the company through its long- and short-term goals. Loans must be paid back. The most effective way to gain credit is through sales.

The best industrial policy is a purchase order. Several years ago, there was a gallium arsenide (GaAs) symposium during which government technical personnel outlined the many military applications which required GaAs chips. At the end of the symposium, a round-table discussion ensued during which the government representatives asked what the government could do to insure a steady supply of high-reliability GaAs chips at reasonable prices. The reply from the President of a small GaAs firm was instantaneous and emphatic: "Give us purchase orders."

Most semiconductors used in defense applications are controlled by Specification Controlled Drawings (SCDs). These documents are based upon standard commercial specifications. They are issued by the DOD after negotiations with the defense contractor. Slight changes to a generic device can create a proliferation of military drawings. The resulting smaller volumes for each drawing reduce the availability of interested suppliers, increase prices and delivery times, and decrease reliability.

The DOD has argued in the past that it needs SCDs to achieve higher performance and reliability. The fact of the matter is that SCDs result in extended delays in the insertion of devices into weapons systems. It now takes from seven to fourteen years for a typical U.S. weapons system to undergo the process from design to deployment. Most semiconductors in these systems represent outdated technology. Furthermore, the reliability level of semiconductors designed into these systems is considerably lower than that of civilian systems. A 1977 report on the screening of military semiconductors revealed that the premium paid by the DOD for high reliability is uneconomic [Hnatek, pp. 101–105, Feb. 3, 1977]. The failure rates for ICs ranged from 6.2 to 50.9% A similar report in 1983 showed that the military still was receiving ICs at a failure rate that would put a private sector company into bankruptcy [Hnatek, pp. 18–24, Jan. 1983]. IC rejects ranged from 8.4 to 16.7%. Japanese and American commercial producers strive for reject rates on the order of 1 part per million.

The DOD now understands that the proliferation of SCDs has created an economic problem. To reduce their number, the DOD introduced in the mid-1980s the Standardized Military/Drawing (SMD) program [DESC, pp. 1–4]. The foreword to the military bulletin on SMDs, issued by the responsible DOD agency, Defense Electronic Supply Center (DESC), states in part:

> The proliferation of industry prepared drawings for the same part used in a variety of military applications has become an ever increasing item of expense to the DOD . . . Standardized Military Drawings (SMDs) are being prepared to eliminate the need for the multitude of contractor prepared drawings for the same device when the minimum requirements for military drawings are sufficient to meet the requirements of the applications on an interim or permanent basis.

The way this works is that a new user-generated parts specification is described in a standard format. Provided that the new specification is not already on file, it is approved and listed in the next issue of the military bulletin on SMDs.

In theory, the SMD program should materially reduce the amount of military drawings. In practice, military design engineers tend to view their requirements as unique and proprietary to their systems. In an environment where cost has not been a major consideration, it is difficult to convince the specification writers that failures cannot be specified or tested out. Reliability is built into the manufacturing process. High yields and reliability are gained by long production runs, not by making a variety of small runs. As chip densities increase and feature dimensions shrink, process control looms larger as the key to device reliability and cost. A recent study by IBM quantified the cost advantages of using commercial-grade products. The study concluded that if electronic components were purchased from commercial sources under existing guidelines, a significant cost savings could be realized without affecting reliability. In the same article that summarized the IBM study, anecdotal evidence was presented for the reliability of commercial-grade computer hardware and software purchased hurriedly for Operation Desert Storm. One Army command unit thanked the supplier for the reliability of its equipment, noting that only four of its 125 personal computers required any repair, and that one PC, blown 50 ft out of a building by a Scud missile, was able to be repaired for normal operation [Burrows, pp. 26–32].

Decontrol Semiconductor Exports

*Shift control of semiconductor device exports from
the DOD to the Department of Commerce.*

Restrict the DOD to controlling the export of semiconductor technology in cases where technology is vital to national defense and is unobtainable from foreign sources. The DOD places too much emphasis on controlling the end product and not enough on design and process know-how [McIntyre, p. 12].

Although it is a more sound business policy to sell product rather than technology, often a company has no choice but to transfer technology through royalty-bearing licenses, or cross-licenses, in order to obtain capital. The effectiveness of technology transfer depends to a great extent upon the ability of the recipient to absorb the technology [Kranzberg, p. 38]. The transfer of semiconductor technology has greatly benefitted U.S. competitors in the marketplace, but not our former military rival, the U.S.S.R. The Soviet Union never had the financial resources, the entrepreneurial experience, the supporting technological infrastructure, or the political system necessary to efficiently absorb illegally acquired U.S. semiconductor technology.

With the decline in the defense market, potential U.S. OEIC suppliers must look abroad. If they are prevented from exporting, the optoelectronic markets will go to Japan by default. Foreign customers often "design out" U.S. products. They cannot depend upon the U.S. as a reliable supplier [Kolcum, p. 89].

Eliminate export controls on semiconductor products that can be obtained from foreign sources. The U.S. intelligence community currently is against the development of a trans-Siberian fiber optic transmission system that will link Europe and Japan [Scully, pp. 21–22]. A Western consortium, including U.S. West, a regional Bell holding company, led in the bidding for the trans-Siberian network, which should take three–five years to complete after construction begins. When it is completed, its technical capabilities will be considerably behind those of the United States.

The design of this communication system calls for a transmission speed of 565 megabits per second (Mb/s). The 17-nation Coordinating Committee on Multilateral Export Controls (CoCom) presently limits the export of optical fiber cable to Russia to speeds of 45 Mb/s and laser diodes to wavelengths of 1000 nm [Haber, pp. 13–14].[10] The present state of the art for operating equipment in the U.S. is 1.2 gigabits per second (Gb/s), twice the proposed speed of the trans-Siberian network. By the time the trans-Siberian network is in operation, transmission rates of 2.0 Gb/s and above will be installed in the U.S. and additional speedier systems will be in the debugging stage. Furthermore, it appears that Russia now possesses the technology to produce long-haul fiber optic cable capable of transmitting at 565 Mb/s ["Soviet low-loss fiber," p. 3].

American intelligence effectiveness will not be affected materially by this project. First, the security characteristics of fiber optic cable are basically the same at 45 or 565 Mb/s. Logically, CoCom should be against the sale of optical transmission systems regardless of the rate of speed.

Second, the major problem facing U.S. intelligence is not collection, but analysis and interpretation. We are receiving too much data, and we need better processing capability.

Third, there are ways to tap into any system, even one that is optoelectronically based. At some point, optics must be converted to electronics. The system can be subverted by existing techniques, well known to the intelligence community, that are designed to pick up and read electrical emissions. Personally, I would prefer that the U.S. design Russian systems. The designer is in the best position to assess its weaknesses and strengths.

The result of denying a U.S.-led consortium to bid on the trans-Siberian project is to ensure the preeminence of European companies in the multibillion dollar Russian communications market. German and French firms now are planning to install fiber optic feeder links in Western Europe, Eastern Europe, and in the Asian countries of the former Soviet Union. These short-haul lower speed links ultimately will be absorbed into the trans-Siberian system.

[10]Effective long-distance fiber optic transmission requires laser diodes operating at wavelengths of 1300 or 1550 nm.

Establish a Technology Database

The ability of Japanese semiconductor firms to acquire and analyze the technical literature of the world is a major factor in their success at commercializing technology. Most semiconductor engineers and scientists in Japan read English-language technical journals. They learn about new technology at the same time as their American counterparts. Very few technically trained Americans read Japanese. They must wait until articles are translated. Many articles are abstracted in English, never fully translated. Others never are translated in any form.

The Japan Information Center of Science and Technology (JICST) is the central organization in Japan for technology and science information [JICST Sales Literature, 1986]. Its main activity is to publish abstracts. It also offers a translation service, an on-line retrieval service, and other services designed to disseminate information. JICST was established in 1957 and is under the control of the Science and Technology Agency, Prime Minister's Office. JICST receives funding from the government and from subscriptions and service fees. JICST started with a capital of $222 000, evenly divided between the government and private enterprise. In FY 1985, the total capital was $98.3 million. Income was $35.4 million: 21% from the government, 34% from subscriptions, 39% from services, and 6% from others.

JICST collects about 12 300 journals annually, of which 47% comes from Japan, 19% from the U.S., 10% from the U.K., 5% from Germany, and 3% from the former Soviet Union. Electrical and electronics journals account for 6.5% of the titles, and physics 3.3%.

The U.S. government should cooperate with the IEEE to organize a collection and dissemination agency similar to that of the JICST. Its first task is the translation of all Japanese journals on electronic and electrical subjects, starting with optoelectronics. Both the government and the IEEE should fund the agency. The latter can add a small charge to its annual subscription fees to cover its contribution. The translation price per page to the user should be nominal, perhaps 5–10% of the market price.

If qualified linguists are not available, the DOD should admit civilians to its language school in Monterey, CA. This facility offers intensive one-year courses of instruction in a multitude of languages. Because the Japanese language is a difficult and time-consuming language to learn, it probably requires a minimum of two years for the needed level of proficiency. A civilian translation and interpretation service skilled in Japanese, Russian, and other scientifically important languages would constitute an inexpensive and effective way to diffuse scientific and engineering information throughout U.S. industry.

Only six U.S. universities offered studies in Japanese culture and technical language in 1989 [Mandell, p. 12]. It is not realistic to believe that any educational program will significantly reduce the imbalance in the data flow between Japan and the U.S. in this century. However, on a long-term basis, it is in the interest of the U.S. government to strengthen its technology base by providing funds for a Japanese language department in all engineering schools that have defense electronics contracts. A coherent well-funded program is necessary in order to seek and acquire competent instructors who should be compensated as part of a defense contract, rather than at standard language department rates.

Not Recommending a National Industrial Policy

The above recommendations are specifically designed to create a technological advantage for the U.S. in the manufacture and marketing of the OEIC, an optoelectronic semiconductor which will make smoother and speedier the merging of computers and communications. A recent report by the National Research Council contrasts the situation in the U.S. "where debate continues over the proper role of the government in industrial and technology policy" with that of Japan where "Japanese government agencies work to create comparative advantages in many fields" [National Research Council, p. 7, 1990].

The first, and most important, recommendation, a national OEIC foundry, may be challenged by those who are against U.S. government participation in manufacturing operations. The fact of the matter is that the federal government has funded semiconductor manufacturing in the past. National Semiconductor has built an $85 million semiconductor facility for the National Security Agency (NSA) dedicated to the production of proprietary chips for NSA applications. The government now is proposing to partially fund the creation of a production facility for high-purity single-crystal silicon material, and to guarantee a minimum annual purchase from this production line [Robertson, p. 1, July 22, 1991].

In a recent survey of nearly 600 Chief Executive Officers of electronic firms, 63% believed that the U.S. should institute a national industrial policy [Rayner, p. 48, Mar. 18, 1991]. I am not proposing a national industrial policy, but only four steps that will allow the U.S. to restore its competitiveness in semiconductors by increasing its chances of becoming the leader in the OEIC market. "The country that commands the three most crucial technologies—semiconductors, computing and communications—will most assuredly command the mightiest industrial bandwagon of the twenty-first century" ["Copycat turns leader?," p. 4]. Semiconductors are the basic elements in computing and communications. Leadership in computing and communications follows leadership in semiconductors. U.S. policymakers must understand that semiconductor technology is a national asset. The U.S. will not be the leading industrial nation in the 21st century if it fails to be internationally competitive in the semiconductor marketplace. A failure to be competitive in the marketplace means that the U.S. semiconductor industry will not generate sufficient profits to fund meaningful R&D. The U.S. government will be faced in the 21st century with the unpleasant choice of either becoming second rate in electronics or infusing massive amounts of cash to restore U.S. competitiveness. The first choice guarantees a chronic trade deficit in electronics. The second choice will be another coffin nail in the budget deficit. Either choice will undermine national security. The U.S. should learn from Japan . . . who learned from IBM . . . *plan ahead.*[11]

---

[11]This was a slogan that IBM adopted sometime after World War II. It was displayed prominently in IBM facilities for employees and visitors to see.

Appendix

TABLE A-1   Estimates of Optoelectronic R&D Expenditures ($ Millions)

	1981	1982	1983	1984	1985	1986	1987
United States							
Industry	65	98	122	192	250	312	375
Government	4	5	6	18	27	27	38
Total	69	103	128	210	277	339	413
Japan							
Industry	106	132	160	206	258	322	386
Government	6	15	21	24	26	22	19
Total	112	147	181	230	284	344	405
Europe							
Industry							
Government							
Total	30	45	65	90	140	165	190
World Total	211	295	374	530	701	848	1008

Source: Tassey, p. 30, Table 2.

TABLE A-2 Corporate R&D Expenditures of Leading Merchant
Semiconductor Producers (U.S. and Japan, 1988)

	R&D as % of Sales	R&D Expenditures ($ Millions)
U.S.		
Motorola (33)*	8.1	665
TI (49)	7.8	494
Intel (80)	11.1	318
National (67)**	11.3	280
AMD (100)	18.5	208
Total	9.4***	1965
Japan		
NEC (19)	8.8	2015
Toshiba (14)	6.4	1841
Hitachi (7)	4.7	2204
Fujitsu (13)	8.9	1505
Mitsubishi (10)	5.5	1019
Total	6.4***	8584

   *Figures in parentheses indicate percentage semiconductor sales to
    corporate sales.
  **National sold its nonsemiconductor operations in 1989.
 ***Total sales divided by total R&D expenditures.

Source: Author's compilation based on: Kelly, pp. 28–29; Rayner and Stallman, pp. 50–58; Stallman and Rayner, pp. 26–42; Instat Inc., p. 30, Sept. 25,
1989; corporate annual reports.

TABLE A-3 Estimated Corporate-Sponsored Semiconductor R&D Expenditures of Leading Merchant Semiconductor Suppliers (U.S. and Japan, 1988)

	Semiconductor R&D as % of Semiconductor Sales	Semiconductor R&D Expenditures ($ Millions)
Motorola (33)*	14.5	399
TI (49)	12.7	395
Intel (80)	16.5	381
National (67)**	17.0	280
AMD (100)	18.5	208
Total	14.7***	1663
NEC (19)	17.5	766
Toshiba (14)	12.5	736
Hitachi (7)	9.0	308
Fujitsu (13)	16.9	600
Mitsubishi (10)	8.8	305
Total	16.4***	2715

*Figures in parentheses indicate percentage of semiconductor sales to corporate sales.

**National sold its nonsemiconductor operations in 1989.

***Total semiconductor sales divided by total semiconductor R&D expenditures.

Source: Author's estimate based on data in: Kelly, pp. 28–29; Rayner and Stallman, pp. 50–58; Stallman and Rayner, pp. 26–42; Instat Inc., p. 30, Sept. 25, 1989.

TABLE A-4 Corporate-Sponsored R&D Expenditures of OEIC Developers
with Merchant Semiconductor Operations (U.S. and Japan, 1988)

	R&D as % of sales	R&D Expenditures ($ Millions)
U.S.		
AT&T (2)*	7.3	2570
Texas Instruments (49)	7.8	494
Motorola (33)	8.1	665
Harris (33)	5.6	117
Hughes (**)	4.9	551
Rockwell (3)	3.6	437
Honeywell (3)	4.5	322
TRW (**)	5.1	356
Westinghouse (**)	5.7	712
Total	6.1***	6224
Japan		
NEC (19)	8.8	2015
Toshiba (14)	6.4	1841
Hitachi (7)	4.7	2204
Fujitsu (13)	8.9	1505
Mitsubishi (10)	5.5	1019
Matsushita (4)	5.6	2280
Total	6.2***	10864

> *Figures in parentheses indicate percentage semiconductor sales to
>   corporate sales.
> **Less than 1% of sales.
> ***Total sales divided by total R&D expenditures.

Source: Author's compilation based on: Kelly, pp. 28–29; Rayner and Stallman, pp. 50–58; Stallman and Rayner, pp. 26–42; Instat Inc., p. 30, Sept. 25,
1989; corporate annual reports.

References

Abegglen, J.C. and Stalk, Jr., G., *Kaisha, The Japanese Corporation*. New York: Basic Books, 1985.

Abbeglen, J.S. and Etori, A., "Japanese technology today," advertisement section in *Scientific American*, Oct. 1980.

"Access Japan," *Business Tokyo*, Feb. 1990.

"After a 12-year wait, Japanese patent awarded," *Defense Electronics*, Dec. 1989.

Agres, T., "Chronic problems at military labs threaten U.S. technology leadership," *Research and Development*, July 1989.

Anamartic Inc., Sales Literature, San Jose, CA, 1989.

"AT&T/Sumitomo joint venture company created," *Lightguide Digest*, issue no. 1, published by AT&T Network Systems, Morristown, NJ, 1989.

Baatz, E.B. and Stallman, L., "Inside the top 50 worldwide chip companies," *Electronic Business*, Apr. 8, 1991.

Bardeen, J., "Solid state physics—1947," *Solid State Technology*, vol. 30, no. 12, Dec. 1987.

Barke, R., *Science, Technology, and Public Policy*. Washington, DC: CQ Press, 1986.

Barrett, J., "House okays R&D grant, loan programs," *Electronic News*, July 22, 1991.

Bell, T., "Optical computing: A field in flux," *IEEE Spectrum*, Aug. 1986.

"Bellcore chief hits drive for fast profits, policy lack," *Electronic News*, May 15, 1989.

Berger, M., "The game could be over in Japan market for U.S. chip-equipment makers," *Electronics*, July 22, 1985.

Berger, M., "High noon for Fujitsu," *Electronics*, May 26, 1986.

Berger, S., Dertouzas, M.I., Lester, R.K., Solow, R.M., and Thurow, L.C., "Toward a new industrial America," *Scientific American*, vol. 260, no. 6, June 1989.

"The billowing beer market," *Business Tokyo*, Feb. 1990.

Bindra, A., "Transmitter IC," *Electronic Engineering Times*, May 16, 1988.

Birkholz, T.A., "Global competitiveness: The six front war," *Solid State Technology*, Apr. 1989.

Blustain, H. and Polishuk, P., "Fiber optics: Technology diffusion and industrial competitiveness," in *Is New Technology Enough?*, D.A. Hicks, Ed. Washington, DC: American Enterprise Institute for Public Policy Research, 1988.

Boyd, J.T., "Compound semiconductor optical waveguides," Special Issue Papers in the Special Issue on Integrated Optics, *Journal of Lightwave Technology*, vol. 6, no. 6, June 1988.

Bradfield, P.L., Brown, T.G., and Hall, D.G., "Electroluminescence from sulfur impurities in a p–n junction formed in epitaxial silicon," *Applied Physics Letters*, vol. 55, no. 2, July 10, 1989.

Burrows, P., "Why the Pentagon must go commercial," *Electronic Business*, June 3, 1991.

Callan, B., "Defense Electronics' top 100 companies," *Defense Electronics*, Jan. 1990.

Carney, J.K., Helix, M.J., and Kolbas, R.M., "Gigabit optoelectronic transmitters," in *5th Annual IEEE GaAs IC Sympo-*

sium Technical Digest, Institute of Electrical and Electronics Engineers, New York, NY, 1983.

Carts, Y.A., "Tiny lasers promise high-speed communications," *Laser Focus World*, Oct. 1989.

"CEOs: U.S. lags in fabs," *Electronic Engineering Times*, Aug. 6, 1990.

Chin, S., "3-D interconnects boost density and performance," *Electronic Products*, Feb. 1980.

Choi, H.K., Turner, G.W., Windhorn, T.H., and Tsaur, B.-Y., "Monolithic integration of GaAs/AlGaAs double-heterostructure LED's with Si MOSFETs," *IEEE Electron Device Letters*, vol. EDL-7, no. 9, Sept. 1986.

"Classified chip-making," *The Institute*, Aug. 1990.

Congressional Budget Office, *The Benefits and Risks of Federal Funding for Sematech*, Sept. 1987.

Congressional Budget Office, *Using R&D Consortia for Commercial Innovation: Sematech, X-Ray Lithography, and High-Resolution Systems*, July 1990.

Connolly, J., "U.S. can't afford semiconductor rescue," *Electronic News*, Dec. 12, 1989.

"Copycat turns leader?," *The Economist*, Aug. 23, 1986.

"Corning and Sumitomo settle patent dispute," *Photonics Spectra*, Jan. 1990.

Costa, B., "Historical remarks," in *Optical Fibre Communication*, Technical Staff of CSELT. New York: McGraw-Hill, 1981.

"Council: More money to U.S. industry," *Electronic Engineering Times*, Sept. 19, 1988.

Council on Competitiveness, *Gaining New Ground: Technology Priorities for America's Future*, Publications Office, Council on Competitiveness, Washington, DC, Mar. 1991.

Dataquest Inc., "Semiconductor market share: Little change at the top," *Electronics*, Aug. 1990.

Defense Electronics Supply Center (DESC), *Military Bulletin-103*, List of Standardized Military Drawings (SMDs), Jan. 8, 1988.

Department of Defense, Office of the Under Secretary of Defense for Acquisition, VHSIC Program Office, *Annual Report for 1987*, Dec. 31, 1987.

Department of Defense, Office of the Assistant Secretary of Defense (Public Affairs), *News Release: MIMIC Phase One Contractors Selected*, May 20, 1988.

Department of Defense, *Critical Technologies Plan*, Mar. 15, 1989.

Department of Defense, *Critical Technologies Plan*, Mar. 15, 1990.

Department of Defense, Office of the Under Secretary of Defense for Research and Engineering (OUSDRE), *Electro-Optics Millimeter/Microwave Technology in Japan*, May 1985.

Dumbleton, J.H., *Management of High-Technology Research and Development*. New York: Elsevier Science Publishers, 1986.

Dummer, G.W.A., *Electronic Inventions and Discoveries*, 3rd ed. Elmsford, NY: Pergamon, 1983.

"Electronics industry finds unlikely support of claims," *Wall Street Journal*, Jan. 6, 1989.

Esaki, L., "Discovery of the tunnel diode," *IEEE Transactions on Electron Devices*, vol. ED-23, July 1976.

Ettenberg, M., "Applications spur diode-laser developments," *Laser Focus World*, May 1991.

Evans, G.A., Carlson, N.W., Hammer, J.M., and Bartolini, R.A., "Surface emitters support 2-D diode-laser technology," *Laser Focus World*, Nov. 1989.

Ferguson, B., "TRW LSI products on the block," *Electronic News*, June 6, 1988.

Flamm, K., *Targeting the Computer*. Washington, DC: The Brookings Institution, 1987.

Forrest, G.T., "OITDA—Japan's E-O information gatekeeper," *Laser Focus World*, Oct. 1989.

Forrest, G.T., "A field of dreams," *Lightwave*, Oct. 1989.

Fouquet, J., "Government support declining for R&D in lasers and electro-optics?," *IEEE Circuits and Devices Magazine*, Sept. 1990.

Fulenwider, J., "Future opportunities in optoelectronics," *Lightwave*, Oct. 1990.

Galatowitsch, S., "DARPA: Turning ideas into products," *Defense Electronics*, July 1991.

Haber, L., "CoCom takes a hard fiber line with East Bloc," *Lightwave*, Mar. 1991.

Harman, T.C., "Narrow-gap semiconductor lasers," in *The Physics of Semimetals and Narrow Gap Semiconductors*, Carter and Bate, Eds. New York: Pergamon, 1971.

Hartman, D.H., Grace, M.K., and Ryan, C.R., "A monolithic silicon photodetector/amplifier IC for fiber and integrated optics applications," *Journal of Lightwave Technology*, vol. LT-3, no. 4, Aug. 1985.

Hartman, D.H., Grace, M.K., and Richard, F.V., "Effective lateral fiber-optic electronic coupling and packaging techniques suitable for VHSIC applications," *Journal of Lightwave Technology*, vol. LT-4, no. 1, Jan. 1986.

Harvard Business School, *SEMATECH: Innovation for America's Future*, 9–389–057, rev. 7/89. Boston, MA: Harvard Business School Publishing Division, 1988.

Hayashi, I., Masahiro, H., and Yoshifumi, K., "Collaborative semiconductor research in Japan," *Proceedings of the IEEE*, vol. 77, no. 9, Sept. 1989.

Hecht, J., "IBM's VLSI optoelectronic chip," *Lasers & Optronics*, Oct. 1989.

Heilbroner, R.L. and Singer, A., *The Economic Transformation of America: 1600 to the Present*, 2nd ed. New York: Harcourt Brace Jovanovich, 1984.

Hnatek, E.R., "High-reliability semiconductors: Paying more doesn't always pay off," *Electronics*, Feb. 3, 1977.

Hnatek, E.R., "The case for component rescreening," *Test & Measurement World*, Jan. 1983.

Honeywell, Letter to Potential GaAs Users, Dec. 11, 1986.

Hooper, L. and Schlesinger, J.M., "Is optical computing the next frontier, or just a nutty idea?," *Wall Street Journal*, Jan. 30, 1990.

Howell, T.R., Noellert, W.A., MacLaughlin, J.H., and Wolff, W., *The Microelectronics Race*. Boulder, CO: Westview Press, 1988.

"Hughes makes fastest InP transistor yet," *Semiconductor International*, Feb. 1989.

Hutcheson, L.D., Haugen, P., and Husain, A., "Optical interconnects replace hardwire," *IEEE Spectrum*, Mar. 1987.

IEEE–USA, *Federal Legislative Agenda*, The Institute of Electrical and Electronics Engineers, New York, NY, 1989.

Ikegami, T. and Kawaguchi, H., "Semiconductor devices in photonic switching," *IEEE Journal on Selected Areas in Communications*, vol. 6, no. 7, Aug. 1988.

Instat Inc., "Top 15 semiconductor suppliers 1988 worldwide sales," *Electronic News*, Sept. 25, 1989.

Instat Inc., "Top 15 merchant suppliers 1988 U.S. semiconductor sales," *Electronic News*, Oct. 16, 1989.

Instat Inc. and Semiconductor Industry Association, "Total semiconductor worldwide," *Electronic News*, Apr. 17, 1989.

"Integrated FO receiver," *Photonics Spectra*, Jan. 1990.

Ishii, T., "Perspectives on the microelectronics industry and technology in Japan," *The Japanese Electronics Challenge*, M. McLean, Ed. New York: St. Martin's Press, 1982.

Iwama, T., Horimatsu, T., Oikawa, Y., Yamaguchi, K., Sasaki, M., Touge, T., Makiuchi, M., Hamaguchi, H., and Wada, O., "4 × 4 OEIC switch module using GaAs substrate," *Journal of Lightwave Technology*, vol. 6, no. 6, June 1988.

Jackson, P. , "Porous silicon emits color," *Lasers & Optronics*, June 1991.

Jackson, P., "IBM's etched mirrors in Zurich," *Lasers & Optronics*, May 1991.

JICST, Japan Information Center of Science and Technology, Sales Literature, C.P.O. Box 1478, Tokyo, Japan, 1986.

Johnston, A.R., Nixon, R.H., Bergman, L.A., and Esener, S., "Optical addressing and clocking of RAM's," *NASA Tech Briefs*, vol. 13, no. 5, May 1989.

Kales, D., "Japan's optoelectronics industry grew 20% in FY 1988," *Laser Focus World*, June 1989.

Kales, D., "Detector industry faces uncertain times," *Laser Focus World*, Nov. 1990.

Kaplan, G., "Business roundup," *IEEE Spectrum*, Jan. 1991.

Karlin, B., "How Japan Inc. is cashing in on free U.S. R&D," *Electronic Business*, Apr. 15, 1987.

Katauskas, J., "Optical computing reaches a crossroads," *R&D Magazine*, Jan. 1991.

Kelly, J., "Three markets shape one industry," *Datamation*, June 15, 1989.

Ketterson, A.A., Tong, M., Seo, J.-W., Nummila, K., Morikuni, J.J., Kang, S.-M., and Adesida, I., "A high-performance AlGaAs/InGaAs/GaAs pseudomorphic MODFET-based monolithic optoelectronic receiver," *IEEE Photonics Technology Letters*, vol. 4, no. 1, Jan. 1992.

Kilcoyne, M.K., Pedrotti, K.D., Beccue, S., and Haber, W., "Optical signal interconnection between GaAs integrated circuit chips," in *Integration and Packaging of Optoelectronic Devices*, D.H. Hartman, R.L. Holman, and D.P. Skinner, Eds., Proceedings of SPIE, 1987.

Kim, M.E., Hong, C.S., Kasemset, D., and Milano, R.A., "GaAlAs integrated optoelectronic transmitter using selective MOCVD epitaxy and planar ion implantation," in *5th Annual IEEE GaAs IC Symposium Technical Digest*, Institute of Electrical and Electronics Engineers, New York, NY, 1983.

Kimura, Y., *The Japanese Semiconductor Industry: Structure, Competitive Strategies and Performance*. Greenwich, CT: Jai Press, 1988.

Kolcum, E.H., "Harris chief warns of economic chaos if manufacturing continues overseas move," *Aviation Week & Space Technology*, Sept. 14, 1987.

Kopin Corporation, *Sales Literature*, Taunton, MA, 1987.

Kranzberg, M., "The technical elements in international technology transfer: Historical prospectives," in *The Political Economy of International Technology Transfer*, J.R. McIntyre and D.S. Papp, Eds. Westport, CT: Quorum Books, 1986.

Lammers, D., "Korean IC makers look to U.S.," *Electronic Engineering Times*, Jan. 1, 1990.

Lee, C.P., Margalit, S., Ury, I., and Yariv, A., "Integration of an injection laser with a Gunn oscillator on a semi-insulating GaAs substrate," *Applied Physics Letters*, vol. 32, 1978.

Lee, T.-P., "Recent advances in long-wavelength semiconductor lasers for optical fiber communication," *Proceedings of the IEEE*, vol. 79, no. 3, Mar. 1991.

Leheny, R.F., Nahory, R.E., Pollack, M.A., Ballman, A.A., Beeber, E.D., Dewinter, J.C., and Martin, R.J., "Integrated In $_{0.53}$ Ga $_{0.47}$ As p-i-n F.E.T. photoreceiver," *Electronic Letters*, vol. 16, 1980.

Levine, B., "Equipment industry facing overhaul," *Electronic News*, section II, May 22, 1989.

Lo, Y.H., Grabbe, P., Iqbal, M.Z., Bhat, R., Gimlett, J.L., Young, J.C., Lin, P.S.D., Gozdz, A.S., Koza, M.A., and Lee, T.P., "Multigigabits 1.5 m/4-shifted DFB OEIC transmitter and its use in transmission experiments," *IEEE Photonics Technology Letters*, vol. 2, no. 9, Sept. 1990.

"Looking at the leaders," *Electronic News*, Sept. 4, 1989.

"Loral agrees to buy defense subsidiary from Honeywell Inc.," *Wall Street Journal*, Nov. 29, 1989.

Makiuchi, M., Hamaguchi, H., Kumai, T., Ito, M., Wada, O., and Sakurai, T., "A monolithic four-channel photoreceiver integrated on a GaAs substrate using metal-semiconductor-metal photodiodes and FETs," *IEEE Electron Device Letters*, vol. EDL-6, no. 12, Dec. 1985.

Malerba, F., *The Semiconductor Business*. Madison, WI: University of Wisconsin Press, 1985.

Mandell, M., "Better ways to watch Japan," *High Technology Business*, June 1989.

Markstein, H.W., "Packaging for high-speed logic," *Electronic Packaging and Production*, Sept. 1987.

Martin, E.W., "Photonics: A lever to change the world," *Photonics Spectra*, Jan. 1991.

McDonald, J.F., Rogers, E.H., Rose, K., and Steckl, A.J., "The trials of wafer-scale integration," *IEEE Spectrum*, Oct. 1984.

McGreivy, D.J., "The role of insulators in VLSI technologies," in *VLSI Technologies Through the 80s and Beyond*, D.J.

McGreivy and K.A. Pickar, Eds. Silver Spring, MD: IEEE Computer Society Press, 1982.

McIntyre, J.R., "Introduction: Critical perspectives on international technology transfer," in *The Political Economy of International Technology Transfer*, J.R. McIntyre and D. Papp, Eds. Westport, CT: Quorum Books, 1986.

Mehler, M., "Semiconductor equipment spending to grow steadily," *Electronic Business*, May 15, 1989.

Menna, R., "GaAs/silicon MMICs," *Microwaves and RF*, Oct. 1987.

Messick, L.J., "Indium phosphide speeds past GaAs," *Electronic Engineering Times*, July 17, 1989.

Midwinter, J.E., "Photonic switching components: Current status and future possibilities," in *IEEE 1987 Topical Meeting on Photonic Switching*, Technical Digest Series, 1987, vol. 13. Washington, DC: Optical Society of America, 1987.

Miller, R.B., "Measurement issues in R&D productivity," in *Understanding R&D Productivity*, H.I. Fusfeld and R.N. Langlois, Eds. New York: Pergamon, 1982.

Miller, S.E., "Integrated optics: An introduction," *Bell Systems Technical Journal*, vol. 48, 1969.

Mirzoeff, J., Ed., *Guide to World Science; Volume 13, Japan*. Guernsey, Channel Islands: Francis Hodgson Limited, 1969.

Mosakowski, P., "Recent trends in start-up activity," *Solid State Technology*, Aug. 1990.

Nagasawa, J. and Forrest, G.T., "Japanese claim to overtake U.S. in worldwide electronics components market," *Laser Focus/Electro-Optics*, Sept. 1987.

Nakamura, M., Suzuki, N., and Ozeki, T., "The superiority of optoelectronic integration for high-speed laser diode modulation," *IEEE Journal of Quantum Electronics*, vol. QE-22, no. 6, June 1986.

Narin, F. and Frame, J.D., "The growth of Japanese science and technology, "*Science*, vol. 245, Aug. 11, 1989.

National Research Council, *Photonics: Maintaining Competitiveness in the Information Era*. Washington, DC: National Academy Press, 1984.

National Research Council, *Science, Technology, and the Future of the U.S.–Japan Relationship.* Washington, DC: National Academy Press, 1990.

National Science Board, *Science Indicators,* 1985.

Nelson, R.R., *High-Technology Policies: A Five-Nation Comparison.* Washington, DC: American Enterprise Institute for Public Policy Research, 1984.

1988 IEEE Annual Report: The Year in Review, The Institute, May 1989.

1990 IEEE Annual Report: The Year in Review, The Institute, May/June 1991.

OFC '90, *Advance Program, Conference on Optical Fiber Communication,* Jan. 1990. Washington, DC: Optical Society of America, 1989.

Office of the U.S. Trade Representative, *Japanese Barriers to U.S. Trade and Recent Japanese Government Trade Initiatives,* Washington, DC, Nov. 1982.

OIDA Brochure, 1991, Optoelectronics Industry Development Association, 400 Hamilton Ave., Palo Alto, CA 94301.

Okimoto, D., "Political context," in *Competitive Edge, The Semiconductor Industry in the U.S. and Japan,* D. Okimoto, T. Sugano, and F. Weinstein, Eds., Stanford University, Stanford, CA, 1984.

Okimoto, D. and Rowen, H., "Chips and defense: What America needs," *The Wall Street Journal,* May 15, 1987.

Oppenheimer, M.F. and Tuths, D.M., *Non Tariff Barriers: The Effects on Corporate Strategy in High-Technology Sectors.* Boulder, CO: Westview Press, 1987.

"Ortel signs agreement with Sumitomo Cement," *Photonics Spectra,* July 1990.

Ota, Y., Miller, R.C., Forrest, S.R., Kaplan, D.R., Seabury, C.W., Huntington, R.B., Johnson, J.G., and Potopowicz, J.R., "Twelve-channel individually addressable InGaAs/InP p-i-n photodiode and InGaAsP/InP LED arrays in a compact package," *Journal of Lightwave Technology,* vol. LT-5, no. 8, Aug. 1987.

Paul, J.K., Ed., *High Technology International Trade and Competition.* Park Ridge, NJ: Noyes Publications, 1984.

"Photonics in Japan: A booming business," *Photonics Spectra*, Sept. 1990.

Pound, R., "Maintain the speed of GaAs in digital systems," *Electric Packaging and Production*, Oct. 1986.

Ray, S., Walton, M.P., Palmquist, S., Hibbs-Brenner, M., and Kalweit, E., "Monolithic optoelectronic circuits for high-speed optical interconnects," in *Integration and Packaging of Optoelectronic Devices*, D.H. Hartman, R.L. Holman, and P.S. Doyle, Eds., Proceedings SPIE 703, 1987.

Rayner, B.C.P., "A blueprint for competitiveness," *Electronic Business*, Mar. 18, 1991.

Rayner, B.C.P., "Foreign firms chip away at the U.S. defense arsenal," *Electronic Business*, Jan. 23, 1989.

Rayner, B.C.P., "Rad-hard chips gain in military tactical market," *Electronic Business*, Sept. 18, 1989.

Rayner, B.C.P. and Stallman, L., "Top 100 R&D spenders increase investment by 15.7%," *Electronic Business*, Aug. 7, 1989.

"Record crowd at OFC reflects surge in sales," *Lightwave*, Mar. 1990.

Reischauer, E.O., *The Japanese Today*. Cambridge, MA: Belknap Press of Harvard University Press, 1988.

Rice, V., "What's right with America's IC equipment makers?" *Electronic Business*, May 15, 1989.

Rice, V., "Chip makers tell DOD: Let us sell rad-hard chips overseas," *Electronic Business*, July 1, 1989.

Richter, A., Steiner, P., Kozlowski, F., and Lang, W., "Current-induced light emission from a porous silicon device," *IEEE Electron Device Letters*, vol. 12, no. 12, pp. 691–692, Dec. 1991.

Robertson, J., "DARPA chief ousted," *Electronic News*, Apr. 23, 1990.

Robertson, J., "DOD would fund silicon line," *Electronic News*, July 22, 1991.

Robinson, B., "Support for opto consortium builds," *Electronic Engineering Times*, Feb. 20, 1989.

Robinson, B., "DARPA taps targets," *Electronic Engineering Times*, May 6, 1991.

Robinson, B., "U.S. opto ICs consortium proposed," *Electronic Engineering Times*, Mar. 6, 1989.

Roessner, J.D., "Technology policy in the United States: Structures and limitations," *Technovation*, vol. 5, Elsevier Science Publishers BV, Amsterdam, 1987.

Rosenblatt, A., "Who's ahead in high-tech?," *IEEE Spectrum*, Apr. 1991.

Rosenthal, D., "Empire built on partnership with workers," *The Atlanta Constitution*, Mar. 30, 1990.

Rothschild, K., "DOD tries to halt loss of key U.S. plants," *Electronic News*, Sept. 14, 1987.

Rotsky, G., "The history of the integrated circuit," *VLSI Systems Design*, Sept. 1988.

Sack, E.A., "Exploding the fabless myth: A little fab can go a long way," *Electronic Engineering Times*, Apr. 22, 1991.

Sah, C.-T., "Evolution of the MOS transistor—From conception to VLSI," *Proceedings of the IEEE*, vol. 76, no. 10, Oct. 1988.

Sakurai, M., "Japan's opto industry soars 20%," *Electronic Engineering Times*, Feb. 20, 1989.

Sanders, III, W.J., "International trade policy," *The Semiconductor Industry*, U.S. Department of Commerce, International Trade Administration, U.S. Government Printing Office, Washington, DC, Apr. 1983.

Santo, B., "DOD funds photonics tech center," *Electronic Engineering Times*, Jan. 29, 1990.

Schneiderman, R., "Federal labs loosen grip on technology," *Microwaves and RF*, Oct. 1988.

Sciberras, E. and Payne, B.D., *Telecommunications Industry*. Harlow, U.K.: Longman Group, 1986.

Scully, S., "U.S. intelligence community oppose trans-Soviet fiber," *Lightwave*, Mar. 1990.

Shandle, J., "AT&T's Koetl savors the challenge of taking on merchant chip giants," *Electronics*, May 1989.

Shibata, J. and Kajiwara, T., "Optics and electronics are living together," *IEEE Spectrum*, Feb. 1989.

Shichijo, H., Lee, J.W., McLevige, W.V., and Taddiken, A.H., "GaAs E/D MESFET 1-kbit static RAM fabricated on silicon substrate," *IEEE Electron Device Letters*, vol. EDL-8, no. 3, Mar. 1987.

Shockley, W., "Negative resistance arising from transit time in semiconductor diodes," *Bell System Technical Journal*, vol. 33, 1954.

Silvernail, L.P., "Optical computing: Does its promise justify the present hype?," *Photonics Spectra*, Sept. 1990.

Sleger, K., Mack, I., Scott, C., and Buot, F., "Compound semiconductor digital integrated circuits," *Microwaves and RF*, Aug. 1989.

"Sony grants $3M for U.S. semicon R&D," *Electronic Engineering Times*, Oct. 1989.

"Soviet low-loss fiber," *Lightwave*, Oct. 1990.

Spicer, W.E., "Organization of the Japanese effort on non-silicon based opto- and micro-electronics," *JTECH Panel Report on Opto- & Microelectronics in Japan*, produced under the Japanese Technology Evaluation Program (JTECH), operated by Science Applications International Corporation, La Jolla, CA, under contract from the U.S. Department of Commerce, May 1985.

Stallman, L. and Rayner, B.C.P., "Vibrant foreign sales can't mask discouraging indicators," *Electronic Business*, July 24, 1989.

Starling, G., "Technological innovation in the communications industry: An analysis of the government's role," in *Communications Policy and the Political Process*, J. Havick, Ed. Westford, CT: Greenwood Press, 1983.

Studt, T., "There's no joy in this year's $150 billion for R&D," *Research and Development*, Jan. 1990.

"Survey warns defense base eroding," *Electronic News*, Nov. 13, 1989.

Suzuki, A., Kasahara, K., and Shikada, M., "InGaAs/InP long wavelength optoelectronic integrated circuits (OEICS) for high-speed optical fiber communication systems," *Journal of Lightwave Technology*, vol. LT-5, no. 10, Oct. 1987.

Tassey, G., *Technology and Economic Assessment of Optoelectronics, Planning Report 23*, National Bureau of Standards, Oct. 1985.

"Technology's top ten," *The Economist*, Aug. 23, 1986.

Tooker, G.L., "Investment and tax policy issues," in *The Future of the Semiconductor, Computer, Robotics and Telecommunications Industries*. Princeton, NJ: compiled by the Editorial Staff, Petrocelli Books, 1984.

Troy, C.T., "IBM scientists develop way to mass-produce, test thousands of semiconductor lasers on a wafer," *Photonics Spectra*, Mar. 1991.

Tsang, D.Z., Smythe, D.L., Chu, A., and Lambert, J.J., "A technology for optical interconnections based on multichip integration," in *Integration and Packaging of Optoelectronic Devices*, D.H. Hartman, R.L. Holman, and D.P. Skinner, Eds., Proceedings of SPIE 703, 1987.

Tsang, W.T., "A bird's eye view and assessment of Japanese optoelectronics: Semiconductor lasers, light-emitting diodes, and integrated optics," JTECH Panel Report on Opto- & Microelectronics in Japan, produced under the Japanese Technology Evaluation Program (JTECH), operated by Science Applications International Corporation, La Jolla, CA, under contract from the U.S. Department of Commerce, May 1985.

U.S. Congress, Office of Technology Assessment, *The Defense Technology Base: Introduction and Overview—A Special Report*, OTA-ISC-374, U.S. Government Printing Office, Washington, DC, Mar. 1988.

U.S. Department of Commerce, International Trade Administration, Office of Telecommunications, *International Competitiveness Study of the Fiber Optics Industry*, 1988.

U.S. Department of Commerce, Technology Administration, *Emerging Technologies*, Spring 1990.

"U.S. dependence on foreign suppliers," *Electronic News*, Dec. 4, 1989.

United States International Trade Commission, *Competitive Factors Influencing World Trade in Integrated Circuits*, USITC Publication 1013, Nov. 1979.

Van Nostrand, J., "Finally Sematech finds a chief exec," *Electronic Engineering Times*, Aug. 1, 1988.

VLSI Research Inc., "The breakdown by product," *Electronic News*, section II, May 22, 1989.

Vogel, S.K., *Japanese High Technology: Implications for Security*, presented at the 1989 Annual Meeting of the American Political Association, Aug. 21–Sept. 3, 1989.

Wada, O., "Optoelectronic integration based on GaAs material," *Optical Quantum Electronics*, vol. QE-20, 1988.

Wada, O., Sakurai, T., and Nakagami, T., "Recent progress in optoelectronic integrated circuits (OEICs)," *IEEE Journal of Quantum Electronics*, vol. QE-22, no. 6, June 1986.

"We are the block," *Electronic Business*, Sept. 4, 1989.

Weber, S., "U.S. industry reacts to a strong message: Change or fail," *Electronics*, Aug. 1989.

Weinstein, F.B., Uenohara, M., and Linvill, J.C., "Technological resources," in *Competitive Edge, The Semiconductor Industry in the U.S. and Japan*, D.I. Okimoto, T. Sugano, and F.B. Weinstein, Eds., Stanford University, Stanford, CA, 1984.

West, L.C., "Picosecond integrated optical logic," *Computer*, vol. 20, no. 12, Dec. 1987.

Wilkins, M., "American–Japanese direct foreign investment relationships: 1930–1952," *Business History Review*, vol. LVI, no. 4, Winter 1982.

Wilson, R.W., Ashton, P. K., and Egan, T.P., *Innovation, Competition, and Government Policy in the Semiconductor Industry*. Lexington, MA: Lexington Books, 1980.

Windhorn, T.H., Turner, G.W., and Metze, G.M., "GaAs/AlGaAs diode lasers on monolithic GaAs/Si substrates," in *Material Research Society Symposium Proceedings*, vol. 67, 1988.

Winkler, E., "Micro Mask agrees to be bought by photoplate supplier for $26 million," *Electronic News*, July 24, 1989.

Yariv, A., "The beginning of the integrated optoelectronic circuit," *IEEE Transcctions on Electron Devices*, vol. ED-31, 1984.

Yust, M., Bar-Chaim, N., Izadpanah, S.H., Margalit, S., Ury, I., Wilt, D., and Yariv, A., "A monolithically integrated optical repeater." *Applied Science Letters*, vol. 35, 1979.

Yoder, S.K., "U.S. firm takes on Japan patent issue," *Wall Street Journal*, Oct. 13, 1988.

Zipser, A., Yoder, S.K., and Schlesinger, J.M., "U.S. chip firms expect little fallout from Texas Instruments patent award," *Wall Street Journal*, Nov. 24, 1989.

Index

128